Advanced Information and Knowledge Processing

Advanced Information and Knowledge Processing

Series editors

Xindong Wu
School of Computing and Informatics
University of Louisiana at Lafayette, Lafayette, LA, USA

Lakhmi C. Jain
University of Technology Sydney, Sydney, Australia

SpringerBriefs in Advanced Information and Knowledge Processing presents concise research in this exciting field. Designed to complement Springer's *Advanced Information and Knowledge Processing* series, this Briefs series provides researchers with a forum to publish their cutting-edge research which is not yet mature enough for a book in the *Advanced Information and Knowledge Processing* series, but which has grown beyond the level of a workshop paper or journal article.

Typical topics may include, but are not restricted to:

Big Data analytics
Big Knowledge
Bioinformatics
Business intelligence
Computer security
Data mining and knowledge discovery
Information quality and privacy
Internet of things
Knowledge management
Knowledge-based software engineering
Machine intelligence
Ontology
Semantic Web
Smart environments
Soft computing
Social networks

SpringerBriefs are published as part of Springer's eBook collection, with millions of users worldwide and are available for individual print and electronic purchase. Briefs are characterized by fast, global electronic dissemination, standard publishing contracts, easy-to-use manuscript preparation and formatting guidelines and expedited production schedules to assist researchers in distributing their research fast and efficiently.

More information about this series at http://www.springer.com/series/16024

Tomasz Wiktorski

Data-intensive Systems

Principles and Fundamentals
using Hadoop and Spark

 Springer

Tomasz Wiktorski
Department of Electrical Engineering
and Computer Science
Faculty of Science and Technology
University of Stavanger
Stavanger, Norway

ISSN 1610-3947 ISSN 2197-8441 (electronic)
Advanced Information and Knowledge Processing
ISSN 2524-5198 ISSN 2524-5201 (electronic)
SpringerBriefs in Advanced Information and Knowledge Processing
ISBN 978-3-030-04602-6 ISBN 978-3-030-04603-3 (eBook)
https://doi.org/10.1007/978-3-030-04603-3

Library of Congress Control Number: 2018962384

This Springer imprint is published by the registered company Springer Nature Switzerland AG
The registered company address is: Gewerbestrasse 11, 6330 Cham, Switzerland

Contents

List of Figures

List of Listings

Chapter 1
Preface

Data-intensive systems are a technological building block supporting Big Data and Data Science applications. Rapid emergence of these systems is driving the development of new books and courses to provide education in the techniques and technologies needed to extract knowledge from large datasets. Current literature lacks centralized learning resources for beginners that integrate administrative, programming, and algorithm design aspects of data-intensive systems.

This book is designed to be a go-to resource for all those beginning to work with data-intensive systems. I focus on explaining the basic underlying concepts, rather than peculiar details of a specific technology or framework. The material in the book is structured following the problem-based approach. It means that content in the chapters is usually focused on developing solutions to simplified, but still realistic, problems using data-intensive technologies and approaches. It is, in contrast, to simply presenting the technologies. I hope such an approach should be more engaging and result in better knowledge retention.

The goal of this book is to take the reader off the ground quickly and then consistently deepen understanding based on a driving example. It would usually require jumping between content in several different books covering various technologies. The focus is on learning a new way of thinking and an approach necessary for big data processing, rather than specifics of technologies that are changing fast. This way the reader gets a solid basis for further steps in the field.

The origin of this book lectures in master's course in data-intensive systems given at the University of Stavanger, Department of Electrical Engineering and Computer Science from year 2013 until now. Some chapters were also a base for guest lectures at Purdue University, College of Information Technology and Lodz University of Technology, Institute of Electronics.

© The Author(s), under exclusive license to Springer Nature Switzerland AG 2019
T. Wiktorski, *Data-intensive Systems*, Advanced Information and Knowledge Processing,
https://doi.org/10.1007/978-3-030-04603-3_1

1.1   Conventions Used in this Book

I use the following typographic conventions in this book:
italic is used to stress new or important terms, filenames, and file extensions;

`constant width` is used for code examples or to refer to parts of code examples, such as variables, function names, etc., and other similar elements in the regular text that should be used literally; and

`constant width italic` is used for parts of code examples that should be replaced by the user-provided values.

1.2   Listed Code

This book takes a hands-on approach to illustrate most of the ideas it presents through numerous listed code examples. You can include parts of the listed code in your own work without permission. In the case that the code is integrated as a part of documentation or any written publication, however, we would appreciate an attribution including title, authors, publisher, and ISBN or DOI. Code is usually provided in Python and Bash. I use these languages because I believe they make it easiest to explain particular concepts for each case. In most cases, examples can be easily translated into different languages.

1.3   Terminology

Data-intensive processing is a type of processing where the primary difficulty is related to volume (and sometimes variety, velocity, or variability) of data. It is in contrast to *computationaly intensive* processing where the main challenge is the amount of computation (i.e., processor cycles). See also Sect. 2.5 for more details.

  Big data is a term typically used in a much wider sense to cover both data-intensive developments and general trends of data usage in organizations.

  Through this book, we will use these terms interchangeably, but typically we will focus on data-intensive with exceptions only when it seems incorrect or unnatural.

1.4   Examples and Exercises

This book stresses the universal applicability of the covered topics to domains that deal with large datasets. Primarily, these domains are big data, data-intensive computing, and also data science. A course designed around this book should be intended as a first step to a variety of roles related to data-intensive systems. The definition and development of the data-intensive systems is ongoing. Two well-established precepts are that data-intensive systems are (1) stimulating the significant development of different approaches to distributed data processing through MapReduce and Hadoop, and (2) refocusing computing domain onto the importance of data rather than simply its computation. This definition guided me in addressing core tasks students and professionals might face in future work or further studies.

The majority of examples in this book are based on Apache Software Foundation Public Mail Archives. I reproduce short parts of this dataset in the book to illustrate concepts and provide sample data for discussion. To perform designed exercises, you can download sample data from the book's website, or if you use Amazon Web Services the archive is available as one of the public datasets. More details are provided later in the book.

To run examples and perform exercises, you will need a running Hadoop installation. You can install Hadoop directly on Linux and Unix systems (including Mac). I recommend, however, that you use one of the available pre-configured virtual machines made available by several Hadoop providers. In my opinion, Hortonworks Sandbox yields itself particularly well to this task. This is because Hortonworks Sandbox stresses remote access from your local machine through SSH and your web browser, exactly the same as real-world Hadoop installations. Alternatively, you can use Amazon Web Services and in particular Elastic MapReduce service. Amazon offers limited free use of its services and provides education grants.

Python is a language of choice for this book, despite Java being the native tongue of Hadoop. In my opinion, code written in Python is more concise and allows to get the core of the algorithm across more efficiently. Nevertheless, you can use Java or C without major problems; Hadoop offers good support for both. Most of Python code is developed using MRJob library. The standard streaming library is also presented at the beginning of the book for reference.

Acknowledgements The work on this book was partially made possible by a grant from Norwegian Center for International Collaboration in Education (SIU)—grant no. NNA-2012/10049. This book was verified and improved through application in several long and short academic courses. Some of these courses were made possible, thanks to grants from the Amazon Web Services Educate initiative.

Chapter 2
Introduction

In this chapter, I explain the importance of data in the modern science, industry, and everyday life. I give examples of datasets that both are large and grow very fast. I also explain hardware trends that drive a need for new paradigms for data processing, which lead to new data processing systems—Data-Intensive Systems. These systems are an essential building block in Data Science application.

I include short descriptions of typical use cases for data-intensive technologies. It is not a complete list, but it roughly covers current technology use, and it should provide you both with good background and inspiration for your practical work with this book.

There are several important frameworks for processing large amounts of data. You can also find many others, typically variations of the major ones, that cover particular niche needs. Differences between frameworks are a result of two major factors: (1) distinct use cases and processing goals, and (2) differences in characteristics of data. The major split is visible between batch and real-time (or stream) processing.

The use cases presented in this chapter typically span this divide. Batch processing is currently most often connected to ETL tasks, but many actual data analyses, such as model building, also happens in batches. A common example of real-time (or stream) processing is time-sensitive data analytic tasks. The border between these two types of processing is fuzzy and constantly shifting. Many frameworks will provide both types of processing, and some attempt to bridge the gap by providing a unified approach to these two types of processing. In addition, with the growing prevalence of in-memory analysis, the distinction becomes less and less significant.

Throughout this book, I predominantly focus on batch processing. Whenever it is possible, I abstract from any particular framework and concentrate on concepts and general technologies, in contrast to specific implementations. I present the fundamentals and expose you to a new way of thinking about data processing. When referring to a particular framework, I typically talk about Hadoop, as it is the most common. All central concepts, however, are usually applicable to the majority of frameworks you might come across, both batch and real-time.

T. Wiktorski, *Data-intensive Systems*, Advanced Information and Knowledge Processing, https://doi.org/10.1007/978-3-030-04603-3_2

2.1   Growing Datasets

Data growth is one of the most important trends in the industry nowadays. The same applies also to science and even to our everyday lives. Eric Schmidt famously said in 2010, as Google's CEO at that time, that every 2 days we create as much information as we did up to 2003. This statement underlines a fact that the data growth we observe is not just linear, but it is accelerating.

Let us look at a few examples of big datasets in Fig. 2.1. You are most probably familiar with a concept of trending articles on Wikipedia. It presents hourly updated Wikipedia trends. Have you ever wondered how these trends are determined? Wikipedia servers store page traffic data aggregated per hour; this leads quite directly to the calculation of trends. However, one caveat is the amount of data it requires. One month of such data requires 50 GBs of storage space and related computational capabilities. While it is not much for each separate month, at moment you want to analyze data from several months, both storage and computation become a considerable challenge. In one of the earlier chapters, we already discussed the analysis of genomic information as one of the major use cases for data-intensive systems, of which 1000 Genomes Project is a famous example. Data collected as a part of this project require 200 TBs of storage.

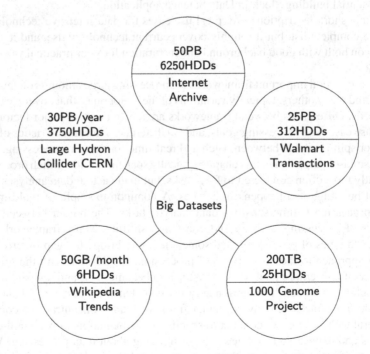

Fig. 2.1 Examples of big datasets. *Source* Troester (2012); European Organization for Nuclear Research (2015); The Internet Archive (2015); Amazon Web Services (2015)

Walmart is one of the major global retailers, so it comes as no surprise that they collect a significant amount of data related to customer transactions; a few years this was estimated at 2.5 PBs. Scientific experiments on Large Hadron Collider (LHC) at CERN generate around 30 PBs of data a year. It is important to notice that this amount of new data has to be stored, backed up, and processed every single year. The last example I would like to offer is Internet Archive, which you are familiar with most probably due to Wayback Machine it offers. It gives you a chance of exploring how websites were changing with time. Currently, total storage used by the Internet Archives reached 50 PBs.

To put these numbers in context, let us consider the largest hard drive available now; it is 8 TBs. It would take over 6000 HDDs to just store Internet Archive; this does not include extra storage to provide redundancy in case of failure. It would take almost 4000 new HDDs every year for LHC data, again in just one copy, without any redundancy.

2.2 Hardware Trends

All these data need to be stored and processed. Storage capacity has been growing quite rapidly in the recent years. At the same time, processing capacity has not been growing accordingly fast. This is easiest visualized by comparing growth in hard drive capacity with hard drive throughput in Fig. 2.2. You can notice that the capacity of a regular hard drive grows exponentially, while throughput linearly. Typically, data generation pace corresponds closely with the increase in storage capabilities, what results in keeping the available storage occupied. The conclusion is that every year it takes longer and longer to read all data from a hard drive. Price and capacity improvements in SSD and memory technologies enable more complex algorithms and frameworks for data processing, but they do not change the general trend.

This brings us to the core of the problem. How can we process all these information within a reasonable time if the capacity growths faster than throughput? One possibility is some form of parallel processing. However, traditional parallel programming techniques separate data processing from data storage. This way reading through data remains a bottleneck. MapReduce paradigm, which is dominant in data-intensive systems, solves this problem combination of map- and reduce-based functional programming routines with a distributed file system. Hadoop is the most common example of such a system. The focus is shifted from moving data to computation, to moving computation to data. This general idea is prevalent in the modern data-intensive; it is a defining feature.

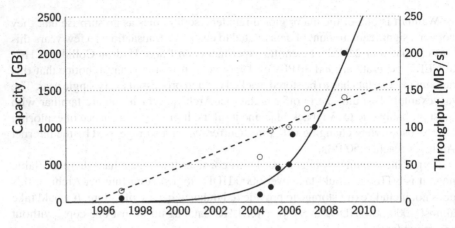

Fig. 2.2 Historical capacity versus throughput for HDDs. *Source* Leventhal (2009)

2.3   The V's of Big Data

Definition of big data through a combination of three characteristics—volume, variety, and velocity—goes back to 2001. In a short paper, Doug Laney (2001) from Gartner introduced for the first time the combination of these three elements as an approach to defining big data. It was later widely adopted and with time extended with many additional adjectives starting with *v*.

The most commonly adopted definition now comes from NIST Big Data Public Working Group, which is one of its documents (SP NIST 2015) defined big data as consisting of "extensive datasets—primarily in the characteristics of volume, variety, velocity, and/or variability—that require a scalable architecture for efficient storage, manipulation, and analysis."

The volume describes the sheer amount of data, variety refers to different types and sources of data, velocity represents the speed of arrival of data, and variability is related to any change in aforementioned characteristics. For instance, average velocity might be low, but it might have picks, which are important to account for. In some case, it is just one of these four characteristics that strongly pronounced in your data, most typically it might be volume. In other cases, none of the characteristics is challenging on its own, but a combination of them leads to a big data challenge.

There are two general technological trends that are associated with big data. The first is horizontal scaling. The traditional way to meet growing demand for computing power was to replace existing machines with new and more powerful; it is called vertical scaling. Due to changes in hardware development, this approach became unsustainable. Instead, a new approach is to build a cluster of ordinary machines and add more machines to the cluster with growing processing need. Such an approach is called horizontal scaling.

The second related trend is related to the software side of the big data. Storage and computation used to be separated that the first step in data processing was sending data

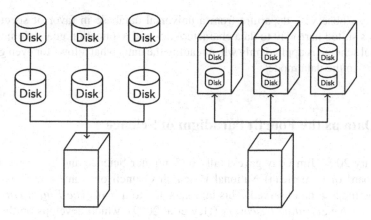

Fig. 2.3 Sending data versus computation

to the computation unit. Again, due to the same changes in hardware development, such an approach became infeasible. New data processing paradigms were necessary to address the necessity to send computation to data. It is presented in Fig. 2.3.

The focus on characteristics of data is ultimately a focus on data themselves. Data become an asset on their own right and a basis of a data-driven organization. The data-centric approach aims to replace simplified models, rules of thumb, and human experts who rely on multi-year experience. The new approach is enabled by the combination of a sufficient amount of data and new data processing paradigms.

2.4 NOSQL

You probably hear the term NOSQL before, even if you were not interested in data-intensive systems. This term as misleading as it is popular. NOSQL started as NoSQL and one letter makes an important difference. NoSQL was to mean the opposite of SQL and was used to collectively describe various databases, which did not adhere to the relational data model. There are two major issues with such an approach.

Relational data model and Structured Query Language are not the same thing. Relational databases can exist without SQL and SQL queries can be executed over non-relational databases. This realization leads to a transformation of NoSQL to NOSQL, which stands for Not Only SQL, but actually means *not only relational*. At least it gives more room for interpretation.

While the name might be misleading, it represents important changes in approach to data modeling and database design. Many sources of data are difficult to fit into the relational model. It might be either unnecessarily complex or simply inefficient for the purpose. New database and storage solutions focus on raw or minimally processed data, instead of normalized data common to relational databases.

You can observe a departure from a universal database in favor of specialized databases suited narrowly to data characteristics and application goals. Sometimes, the actual schema is applied only when reading the data, what allows for even greater flexibility in the data model.

2.5   Data as the Fourth Paradigm of Science

In January 2007, Jim Gray gave a talk to Computer Science and Telecommunications Board of (American) National Research Council on transformations in the scientific method he observed. This talk later led to a book *The Fourth Paradigm. Data-Intensive Scientific Discovery* (Hey et al 2009), which develops on the ideas presented in the talk.

Gray groups history of science in three paradigms. In the beginning, science was *empirical* (or *experimental*), and it revolved around describing and testing observed phenomena. It was followed by *theoretical science*, in which generalized models were created in form of laws and theories. With time, phenomena we wished to describe became too complex to be contained in generalized models. This gave rise to *computational science*, focused on simulation, which was assisted by exponential growth in computing power—Moore's Law. Relying on the growth of computing power of individual desktop machines, and also supercomputers, is referred to as vertical scaling. Timescale between science paradigms is presented in Fig. 2.4.

Currently, we observe an emergence of a new fourth paradigm in science, *data-intensive*. Exponentially increasing amounts of data are collected, for instance, from advanced instruments (such as Large Hadron Collider at CERN), everyday sensors (Internet of Things), or wearable devices (such as activity bracelets). Data become central to the scientific process. They require new methodologies, tools, infrastructures, and differently trained scientists.

Gray also observes that in data-intensive science problems and methods can be abstracted from a specific context. Usually, two general approaches are common "looking for needles in haystacks or looking for the haystacks themselves." There are several generic tools for data collection and analysis, which support these approaches.

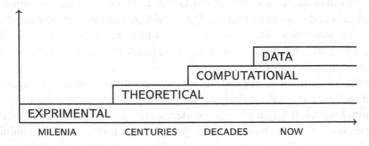

Fig. 2.4 The timescale of science paradigms

At the same time, exponentially growing storage capacity—Kryder's Law—enables preservation of collected data for analysis. However, computing power did not continue to develop as it did in earlier decades. In order to handle the increasing amount of data, while computing power does not increase accordingly, new computing architecture is commonly adopted. It relies on parallel processing on a set of independent machines, where new machines are added when the current cluster of machines becomes insufficient. Such an approach is referred to a horizontal scaling.

2.6   Example Applications

The trends in science and industry confirm the importance of data-intensive approach. In this section, I shortly describe typical use cases for data-intensive technologies. This is not a complete list, but it roughly covers current technology application, and it should provide you both with good background and inspiration for your own practical work with this book. I shortly describe each use case in general and then provide a few short examples.

2.6.1   Data Hub

One of the most common use cases for Hadoop and other similar frameworks is as a next-generation data hub or data warehouse. This applies to both enterprise and scientific data. The tasks can include storage, cleaning, transformation, analysis, and archiving of diverse data. Almost any data-intensive system uses some type of batch processing framework for archiving and ETL tasks, while analysis is sometimes performed outside of the basic framework. This is achieved either by providing special functionality tools, like Hive, that are simply an extension of the basic framework, or by a separate framework, like Spark, that can integrate itself with the batch layer.

Data hub can be seen as an umbrella use case for many others, which are described further in this chapter. It does not, however, limit its importance. The overlap between the use cases that data hub represents provides the necessary weight to ensure the resources for developments that result in higher quality tools.

Facebook can be seen as a good example of adopting data-intensive technologies into its data flow. After years of attempts to tweak MySQL to meet the demand of a growing user base, Facebook now bases many of its operations on two Hadoop clusters that together provide 15PB of storage, spread among 1400 machines. These clusters provide reporting, analytics, and machine learning functionality for many of Facebook's services[1]. You can find more users of Hadoop on Hadoop's Powered By website https://wiki.apache.org/hadoop/PoweredBy.

---

[1] Information current as per March 2016

Facebook is also famous for open-sourcing its data processing hardware architecture. This included custom-designed servers, power supplies, server risks, and battery backup systems. You can find more information in the Open Compute Project website https://www.opencompute.org.

The CMS Tier-2 center at Purdue University is a good example of Hadoop's usefulness for data archiving in the scientific domain. The CMS Tier-2 center is based on Hadoop 2 and provides over 4PBs of storage for the Compact Muon Solenoid experiment conducted at the Large Hadron Collider at CERN[2]. This infrastructure is used to support the computational needs of physicists primarily in US, but also around the globe.

The main difference that modern data-intensive frameworks bring is the focus on making data available for processing. This breaches the earlier gap between archiving and high-performance analysis. This disappearing gap shifts the technological focus from reporting to facilitating advanced analysis, which consequently leads to increased data collection and consumption. You can also notice a growing use of the term data hub instead of data warehouse. While there is little inherent difference between the two terms, this terminology shift is a confirmation of the conceptual and technological shifts happening in the background.

2.6.2   Search and Recommendations

Search is possibly one of the first applications of data-intensive technologies. Notably, the MapReduce approach as applied to data-intensive processing has originated from Google's work on processing its search index. Specific technologies and approaches might differ between companies, but search and recommendations, as we know them today, are dependent upon processing large datasets.

Google made a breakthrough in search quality by effectively indexing "everything" on the Internet by pioneering the use of data-intensive technologies and from important insights on how the quality of websites can be measured. While the principles of the many algorithms that Google used might have been known for decades, it was the scale and automation of the processes that brought a qualitative change. The specific technologies used for processing their search index have developed since then, but the principle has remained. Other companies followed suit, such as LinkedIn with recommendations on "people you may know", all based on the analysis of multiple professional networks.

Amazon is famous for its suggestion of similar and related products. This service relies on analyzing your shopping patterns and correlating them with all other customers. They take advantage of having control over all your past shopping, logged under easily identifiable accounts, using data-intensive technologies that allow them

---

[2]Information current as per October 2015

to quickly connect your information with others' shopping patterns to provide you with suggestions that might increase their profit and possibly your satisfaction (at least short-term).

2.6.3 Retail Optimization

Ability to correlate data from many sources, such as from credit card or general POS transactions, location, or social media, is of great interest to many brick-and-mortar and online retailers. Many companies you may not associate with advanced ICT operations are, in fact, very active in the domain. Their profit margins might often depend on the successful application of large-scale data analysis.

One of the most highly profiled cases of the application of data analysis to shopping patterns is associated with Target, partially due to individual emotional impact. Several years ago, Target noticed that shopping habits on one of its customers started changing. Comparing her with other customers, they concluded that the person was pregnant and started sending her coupons to pregnancy-related items. The Target card, however, happened to belong to the customer's father, who was unaware of his daughter's pregnancy until he started receiving the coupons. This just goes to show that the results of data analysis should be applied very carefully. If you are interested, you can read more about this case and targeting customers in New York Times article—How Companies Learn Your Secrets https://www.nytimes.com/2012/02/19/magazine/shopping-habits.html.

While Walmart and Nordstrom have never had such a high-profile case, they both run well-known data labs that focus on research and development in the area of data-intensive analysis. The goal is to improve various metrics, such as product availability, waste reduction, customer profiling, etc.

2.6.4 Healthcare

Various sub-domains of health care and medicine rely on processing large quantities of data, even if this is not obvious at the first glance. These sub-domains cover a wide range of applications, from pharmaceutical research to DNA analysis, modeling of neural signals, to self-monitoring devices like activity bracelets. This shows how pervasive data is in a rather traditional discipline—one we mostly associate with the sporadic doctor's visit.

Let us look at the application of data to DNA analysis. DNA is first serialized as a long series of characters. From a computer science perspective, the problem of DNA analysis is then reduced to pattern recognition in strings. To give some perspective on the scale of the problem, let us take a look at a human genome. The human genome has about three billion nucleotides (characters), which is roughly the same as English Wikipedia. The genome is analyzed in subsequences, typically generating up to 100 GB for a single genome. Data-intensive technologies answer many of the technical challenges, providing ground for medical research. You can read more about using Hadoop for genome analysis on Cloudera's blog (Christophe Bisciglia 2009).

Self-monitoring devices have become very common in the last few years. The most common are bracelets, and more recently also watches. A crucial part of their functionality is provided through the applications, typically mobile, that collect the data from the bracelets or watches and provide various self-monitoring functions. Additionally, you can typically compete with your friends, providing important social stimuli to improve goal achievement. To achieve all these functionalities, all data is stored centrally by the bracelet's producers and sometimes also by third-party services. It is often far more data than you would expect. What you see in the app is typically only a summary, and most of the producers store data with far greater granularity, usually with ca.5 min intervals. The simple statistics you see in the app do not yet reveal how this data might be used in the future.

2.6.5   Internet of Things

In recent years, we have seen an increase in the amount and data resolution of sensors and other connected devices. This leads to an increased demand for data storage and computation capabilities to extract value from the produced data.

SEEDS (2018) was a project working on self-learning energy-efficient buildings. The idea was based on collecting enough data to automatically construe accurate building models as an alternative to hand-constructed models, which are often only approximate. Data was collected from sensors, including presence, temperature, air quality, lightness, and humidity. To make data available for self-learning and optimization algorithms, and for validation after almost 4 years of the project, OpenTSDB was used as a data archive. OpenTSDB is a distributed time series database used for collecting metrics from data centers and sensors. Another example of a project using similar technology is BEACON (2018) at UC Berkeley, which collects CO2 levels in Bay Area in the USA.

Smart Cities become one of the most visible IoT applications for data-intensive technologies because extracting insights from data is a common theme among Smart City strategies. Smart Cities can be understood as the collaboration between and improvement of city services, based on data-driven decisions and processes. The reliance on data-driven decisions is also why Smart Cities become consumers of data hub technologies, as described earlier in this chapter. Smart Cities experience growing quantities of data, not only from an increase in the amount of sensor-generated data but also because data from previous years are maintained in the hub. It allows new services and long-term impact monitoring, but also puts requirements on data handling infrastructure.

Another interesting use of Internet of Things-type systems is agriculture. Through the use of distributed sensor networks, large farms can monitor the individual growing conditions for each plant. Data is centrally collected and analyzed in conjunction with additional information, such as weather forecasts and past growth data, to devise strategies that increase yield from the field. An example of such a system is DataFloq (2018) from John Deere, which is a holistic platform that offers sensors, collection, and storage infrastructures, and basic analytic functionality. Precise weather data can also be obtained from services like Weather Terrain (2018). The major increase in farms using sensors is definitely a data challenge, but it also opens new areas of data exploration for agriculture.

2.7 Main Tools

This section provides a very short overview of the two main tools for data-intensive processing, which we consider in this book. The goal is to help you to develop a general intuition about the landscape of available tools, so you are able to place the coming chapters in a wider context. It is particularly important for the next chapter, in which you right away dive into programming a Hadoop system.

2.7.1 Hadoop

Apache Hadoop is currently the main open-source system for handling large volumes of data. It consists of two main parts: Hadoop Distributed File System (HDFS) and MapReduce, which provide support for storage and processing, respectively. Hadoop is a de facto industrial standard, and commercial support is provided mainly by Cloudera and Hortonworks. Hadoop is integrated into other commercial data platforms, such as IBM Big Insights.

Hadoop gave a beginning to many projects that extend its functionality. Two of them include HBase—a distributed column-oriented database (an example of a NOSQL database) that uses HDFS as a storage medium; and Hive—a data warehouse framework with SQL-like query language.

2.7.2 Spark

Apache Spark is a data-intensive processing framework that makes use of the distributed memory of nodes in the computing cluster to provide high-performance processing. It does not replace Hadoop, but rather answers many of its shortcomings. This includes reduced overheads, thanks to better JVM management, a wider range of processing routines beyond the map and reduce, improved support for complex algorithms like those common to machine learning. It still relies on HDFS, another file system, or database for permanent storage.

It is fast becoming a go-to tool for rapid prototyping, implementing data science workflows, machine learning at scale, and stream processing, while Hadoop remains the standard for established ETL pipelines.

2.8 Exercises

Exercise 2.1   Identify and inspect public datasets that fit each of the presented examples. Where does the data come from, how can you access it, and how much data is available (time span, size, etc.)?

Exercise 2.2   Extend the graph from Sect. 2.2 with the current data. Is the trend changing?

Exercise 2.3   Extend the graph from Sect. 2.2 with two new curves (both current and historical data), one for memory and one for SSDs. How can you describe these two new trends? Do they correlate with the development of new data-intensive frameworks, such as Spark?

Exercise 2.4   Consider services you use every day, such as Facebook or Spotify. Investigate what data-intensive technologies they are using.

References

Troester M (2012) Big data meets big data analytics. SAS Institute Inc., Cary, NC

European Organization for Nuclear Research. Computing/CERN (2015) http://home.web.cern.ch/about/computing (visited on 06/18/2015)

The Internet Archive. Internet Archive: Petabox (2015) https://archive.org/web/petabox.php (visited on 06/18/2015)

Amazon Web Services. Public Data Sets on AWS (2015) http://aws.amazon.com/public-data-sets/ (visited on 06/18/2015)

Leventhal A (2009) Triple-parity RAID and beyond. Queue 7(11):30

Laney D (2001) 3D data management: controlling data volume, velocity and variety. In: META Group Research Note vol 6, p 70

SP NIST. 1500-1 NIST Big Data interoperability Framework (NBDIF): Volume 1: Definitions, September 2015 http://nvlpubs.nist.gov/nistpubs/SpecialPublications/NIST

Hey AJG, Tansley S, Tolle KM et al (2009) The fourth paradigm: data-intensive scientific discovery, vol 1. Microsoft Research Redmond, WA

Christophe Bisciglia. Analyzing Human Genomes with Apache Hadoop. en-US (2009) http://blog.cloudera.com/blog/2009/10/analyzing-human-genomes-with-hadoop/ (visited on 10/02/2018)

Self learning Energy Efficient builDings and open Spaces/Projects/FP7-ICT. en https://cordis.europa.eu/project/rcn/100193_en.html (visited on 10/02/2018)

BeACON > Home http://beacon.berkeley.edu/ (visited on 10/02/2018)

John Deere Is Revolutionizing Farming With Big Data. en-US https://datafloq.com/read/john-deere-revolutionizing-farmingbig-data/511 (visited on 10/02/2018)

Farm Weather App for Agricultural Weather/aWhere. http://www.awhere.com/products/weather-awhere (visited on 10/02/2018)

Chapter 3
Hadoop 101 and Reference Scenario

This chapter serves as a 101 guide to get you up and running with basic data-intensive operations using Hadoop and Hortonworks Sandbox in less than one hour. I have made a small sample of a dataset available so that you can efficiently run the examples without a bigger cluster. I want to show that you can use all of the basic technologies almost immediately. Presented examples are purposefully very simple to provide you with a basic working skeleton application that will reduce entrance barriers for more advanced topics. You will explore these topics in later chapters that will be based on the code and dataset you will become familiar with right now.

Some information in this chapter might change slightly with new versions of the software.

3.1 Reference Scenario

First, I want to introduce a reference scenario that I will follow through the book. Imagine, you have large email archive and you answer questions about it. You need to find a way to answer these questions with the data you have. This will require a variety of tools, and I will show several of them throughout this book. Even if some questions seem easy, data might not be in a format that is easy to work with. Even if the format is right, the amount of data might create a challenge. I use Apache Software Foundation Public Mail Archives to illustrate all exercises in this book. Questions I am interested in are which organizations are the biggest contributors to Apache projects? There are many ways to answers this question, depending on what is meant by biggest contributor. I will answer this question by analyzing the email domains used for posting in the mailing lists. Of course, there are good reasons to think that the answer I provide this way is not fully precise; however, it (1) is a reasonable estimate, and (2) illustrates how the technologies can be applied, which is the purpose of this book.

© The Author(s), under exclusive license to Springer Nature Switzerland AG 2019   19
T. Wiktorski, *Data-intensive Systems*, Advanced Information and Knowledge Processing,
https://doi.org/10.1007/978-3-030-04603-3_3

The whole dataset has 200GB, and it is easy to use subsets of it. Different subsets of different sizes are available as an easy download from the book's website. The dataset contains emails from mailing lists from over 80 Apache Software Foundation projects as of July 11, 2011. Further descriptions can be found on `aws.amazon.com/datasets/7791434387204566`; you can browse mail archive contents on `mail-archives.apache.org/mod_mbox/`. This dataset provides a useful combination of structured and unstructured data, which I will employ to explain different types of Hadoop use cases and tools. The dataset contains information about senders, recipients, delivery paths, sending and delivery times, subjects, body, thread topics, etc. The first few fields provide a basis for analysis of structured data and the last two of unstructured data. Sometimes, the best way to proceed is to combine both approaches. I will employ each of them in this chapter and in the following chapters. Be aware, though, that this dataset poses a common problem: raw data available in the snapshots are not divided by email but are a stream of emails.

This chapter is self-contained. All code and instructions provided can be implemented directly without any additional work or knowledge (except basic programming abilities and a working Hadoop cluster). I explain shortly each operation you should perform, but avoid most of the details. In case you find yourself wondering about rationale and specifics of particular operations, be patient, I will come to it in the following chapters. This chapter is just a taste of what is coming. I am trying to make you start working with real systems, even if you are not yet fully read.

All the code in this chapter (and also rest of the book) is primarily written in Python. While Python is not native to Hadoop, you might find it simpler and more convenient. Ultimately, the choice of language does not have any significant effect on the content presented in this book. All the examples could also be coded in Java, C, or other languages. The presented examples are based on Hortonworks Sandbox, but you can use them without any major changes on AWS Elastic MapReduce and Elastic Hadoop on OpenStack.

The dataset `hadoop.txt` we are using is partially preprocessed. Originally, the Apache Mail Archives come as a large set of compressed files that cannot be analyzed immediately. In this case, the preprocessing was minimal, and files were extracted from compressed archives and simply merged in one large file. The preprocessing is an interesting problem in itself, and Hadoop is a very useful tool to solve it.

For the structured data analysis, we are using `hadoop.csv`. This dataset required more preprocessing to extract relevant fields and put them into a more structured format.

There are also two other helper datasets that come with this book. `Hadoop_1m.txt` contains simply first one million lines from `hadoop.txt`, it is helpful when testing the algorithms, especially when working with suggested exercises in several chapters. Due to its smaller size, it will allow you to test your code much faster.

Another dataset `email_id_body_1m.txt` contains one million lines from the output of algorithm presented in Sect. 7.4. It is useful for testing the algorithm in the Sect. 7.5. It allows you to perform `inline` and `local` testing before executing an algorithm on Hadoop with data in HDFS.

3.2 Hadoop Setup

There are three Hadoop setups that I recommend you try while learning from this book. They are not exclusive. My recommendation is that you, in fact, use them all continuously while working with this book. First is a simple Docker container from Hortonworks called Sandbox. You can run it on your own computer, which makes it a perfect solution for initial testing of your code. It is also available as a virtual machine. Hortonworks Sandbox and alternatives like Cloudera Quickstart VM are far easier and cleaner solution than installing Hadoop in a standalone mode.

Second is Amazon Web Services Elastic MapReduce (AWS EMR). It is probably the simplest way to get a multi-node Hadoop cluster. AWS offers education and research grants, which make its use essentially free in a classroom and research setting. Unfortunately, AWS EMR is not included in the AWS free tier usage. I recommend that you only start deploying your code on AWS EMR when it has already been tested extensively on Sandbox.

The third setup is based on OpenStack. It provides you with Amazon-like Cloud environment, which you can create yourself on your own hardware. If you are following this book on your own, outside of a classroom setting, you might want to skip over this particular setup. It requires extra time and hardware, which you might not necessarily have. However, if you are interested in Cloud system, you should definitely try it.

In addition to these three Hadoop setups, there are several ways to run your job depending on the language you write in and the required distribution mode. We will talk about different alternatives later in the book. In this chapter, I will simply show you two different ways to execute jobs written in Python. The focus is on using a library called MRJob, which simplifies the execution of jobs written in Python. For reference, I will also execute such jobs using Hadoop Streaming, which is a standard method for jobs not written in Java or C.

To start working with Hortonworks Sandbox, you need the Sandbox[1] itself, and virtualization software such as Docker[2] or Oracle VirtualBox.[3] The recommended configuration is 4 CPUs and 8 GB of RAM; however, 2 CPU and 6 GBs of RAM should be enough for the purpose of following this book. Reducing the amount of RAM will make some of the services unavailable. In principle, Docker version seems

---

[1] http://hortonworks.com/products/hortonworks-sandbox/

[2] https://www.docker.com/

[3] https://www.virtualbox.org

to put less load on the host machine and I recommend it. Follow instructions on the Hortonworks website to install Sandbox on Docker.

Listing 3.1 Hortonworks Sandbox setup

```
1 #start Sandbox
2 docker start sandbox-hdp
3 docker start sandbox-proxy
4
5 #from local machine
6 ssh root@127.0.0.1 -p 2222
7
8 #from Sandbox VM
9 yum -y install python-pip
10 pip install mrjob==0.5.10
```

To start Sandbox in the container, you need to issue commands in line 2 and 3 from Listing 3.1. It might take several minutes or longer to start all services. Right after the installation, container will be started automatically. To check if all services have started or to diagnose any issues, you can look at Ambari available at http://127.0.0.1:8080/, with password and user maria_dev. Ambari, in Fig. 3.1, shows the status of each service and displays useful metrics, in addition to giving you a possibility to perform the most important actions like start and stop.

There are many operations you can perform through web interfaces, such as Ambari. To get down to programming, however, you will usually have to gain access to the container (or a remote machine) through SSH. This way will also apply to other Hadoop distributions and setups. In general, this method allows you to trans-

Fig. 3.1 Ambari in Hortonworks Sandbox

fer your code and skills between different environments, such as Sandbox, AWS, and OpenStack, with minor adjustments. To access the container with the Sandbox, you have to SSH to a localhost, as in Listing 3.2 line 6. The password to the virtual machine is `hadoop`; you will be asked to change the password the first time you log in.

Sometimes, if you have used an older version of the Sandbox, host identification might have changed preventing you from being able to login to the virtual machine through SSH. You should then remove the old identification in the SSH known hosts file. Just Google the location of the file for your operating system. On Mac, you should find it at `Users/your-user-name/.ssh/known_hosts`.

We start by installing some extra software in the Sandbox container in Listing 3.1. After connecting to the Sandbox in line 2, we install `pip` in line 5. In some versions, it might already be included. You can verify it, by simply running `pip` command. If it is already installed, you can skip the step at line 5. Finally, we install MRJob in line 6. It is important to install the right version. At the current moment, I recommend version 0.5.10. Newer version requires a higher version of Python than the one installed on Hortonworks Sandbox. Upgrading the main python installation in CentOS is not recommended, because it is one of the key system components with many dependencies. If you need a newer Python version, you should instead install it as an alternative to the main version. In such a case, you also have to make sure that pip is tied to the Python installation you want to.

Listing 3.2 Accessing Hortonworks Sandbox and copying data

```
1 #from local machine
2 scp -P 2222 -r dis_materials root@127.0.0.1:~
3
4 #from Sandbox VM
5 hadoop fs -mkdir /dis_materials
6 hadoop fs -put dis_materials/*.txt dis_materials/*.csv /dis_materials
```

Finally, you can also copy example data and code to the cluster. You can download them from the book's website. This is presented in Listing 3.2. You use SCP command to copy the data in line 2. Please notice that the scp, in this example, runs from the local machine. If you have before sshed to the cluster, you should terminate that connection first. After the data and code are copied, SSH again to the cluster and copy data from the local drive on the cluster to HDFS, lines 6–7. HDFS is the very basis of data-intensive processing in Hadoop; I will discuss it in details later in the book.

3.3 Analyzing Unstructured Data

In this section, I present you with a simple analysis of unstructured data. Take a look at Listing 3.3, which contains a sample from a hadoop.txt, the file that is a part of materials available for this book. It contains thousands of such mostly unstructured emails. It is possible to deduce some structural elements. I take advantage of them for analysis and also for prepping a CSV file for structured data analysis in the next section. I will extract the sender's domain from all these emails and count the frequency of each domain. This will help us answer the question of which organizations are most active in contributing to the Hadoop development. I will utilize Hadoop Streaming, which allows us to write Hadoop programs in almost any language; in my case, the language is Python.

Listing 3.3 Sample content from hadoop.txt file

```
 1  MIME-Version: 1.0
 2  Received: by 10.231.149.140 with SMTP id
        t12mr14510345ibv.100.1286396237458;
 3   Wed, 06 Oct 2010 13:17:17 -0700 (PDT)
 4  Sender: jdcryans@gmail.com
 5  Received: by 10.231.19.137 with HTTP; Wed, 6 Oct 2010 13:17:17 -0700
        (PDT)
 6  Date: Wed, 6 Oct 2010 13:17:17 -0700
 7  X-Google-Sender-Auth: ACRzZXto3gBOzxrC-iLr_wFT94s
 8  Message-ID:
        <AANLkTinj-MOcTLZvg3k=Zha3GHwp68SvH1m3p9XSYWm=@mail.gmail.com>
 9  Subject: [ANN] HBase-0.89.20100924, our third 'developer release',
        is now
10   available for download
11  From: Jean-Daniel Cryans <jdcryans@apache.org>
12  To: general@hadoop.apache.org, user@hbase.apache.org
13  Content-Type: text/plain; charset=ISO-8859-1
14
15  The HBase Team would like to announce the availability of the third
        in our 'developer release' series, hbase-0.89.20100924. Binary
        and source tar balls are available here:
16
17   http://www.apache.org/dyn/closer.cgi/hbase/
18
19  You can browse the documentation here:
20
21   http://hbase.apache.org/docs/r0.89.20100924/
22
23  This release contains a rollback of some features that were added to
        the master in 0726 that made it more unstable, but also contains
        about 40 fixes for bugs (including more reliable region server
        failovers) and performance improvements.
```

A basic MapReduce program flow is depicted in Fig. 3.2. First, all data runs through map phase. This operation is automatically parallelized for performance. This means that, in fact, the Hadoop system automatically splits the input data into several parts and applies map function to each part independently. Then, these partial results are fed to the reducer function that combines them in one complete piece.

How this particular mapper performs the counting is not that important right now. It is important that many mappers are automatically executed, each processing independent part of the input. The same applies to reduce. The most important part

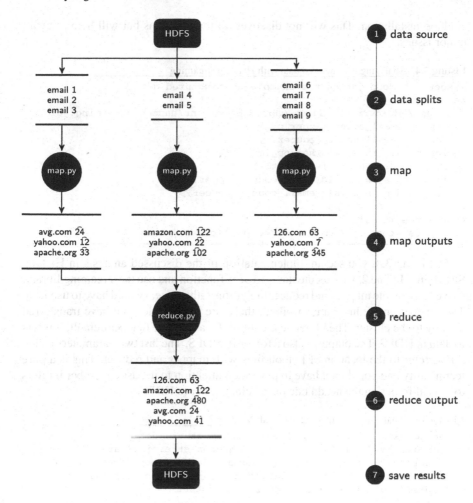

Fig. 3.2 Simple MapReduce workflow overview

is that it reliably combines partial results from all the maps into the final result. This last operation is, in principle, done without parallelism to ensure consistency of data. See further chapters for details and exceptions.

Now that you have a general idea of how the processing is performed, let us code it. I will show you how to run the code using MRJob and Hadoop Streaming. For Hadoop Streaming there might be differences regarding the location of Streaming library between different distributions, so you might have to modify that path in your case. Such minor differences are common for different distributions of open-source software. You will see that in production systems, where you often do not have a choice of distributions, you simply have to work with what you get. I will also show you how you can easily test your code using a regular command line without any

Hadoop installation. This will not discover all the problems but will help to avoid major issues.

Listing 3.4 Analyzing unstructured data with Hadoop streaming

```
1 #execute hadoop job on the example unstructured data
2 hadoop jar
       /usr/hdp/current/hadoop-mapreduce-client/hadoop-streaming.jar \
3   -mapper email_count_mapper.py \
4   -reducer email_count_reducer.py \
5   -input /dis_materials/hadoop_1m.txt \
6   -output /dis_materials/output1 \
7   -file ~/dis_materials/email_count_mapper.py \
8   -file ~/dis_materials/email_count_reducer.py
9
10 #check the result
11 hadoop fs -text /dis_materials/output1/part* | less
```

In Listing 3.4, you see an implementation of the discussed analysis in Hadoop Streaming. In line 2, we execute the hadoop function and use the streaming library, so we can execute mapper and reducer in Python. Otherwise, I would have to use Java. The mapper and reducer are specified; these are Python files you have transferred already to the cluster. Then I define the input. Please notice that per default, it refers to data in HDFS. The output is also a folder in HDFS. The last two parameters called file refer to the location of Python files with mapper and reducer. This is a pure technicality and you do not have to pay much attention to it. Just remember it refers to a local folder on the head node of the cluster, not to HDFS.

Listing 3.5 Analyzing unstructured data with MRJob

```
1 #execute hadoop job on the example unstructured data
2 python dis_materials/count_sum.py --hadoop-streaming-jar
       /usr/hdp/current/hadoop-mapreduce-client/hadoop-streaming.jar -r
       hadoop hdfs:///dis_materials/hadoop_1m.txt --output-dir
       hdfs:///dis_materials/output2 --no-output
3
4 #check the result
5 hadoop fs -text /dis_materials/output2/part* | less
```

In Listing 3.5, you see the implementation of the discussed analysis in MRJob. In line 2, we execute a python script that contains all mapper, reducer, and combiner in one file. We specify a so-called runner, which in this case is Hadoop. Then I define input and output. Please notice that in this case, you have to specify that the file is in HDFS, both for input and output. The last parameter -no-output tells MRJob not to display output to STDOUT, which is a default behavior.

When you execute the code, Hadoop will start map jobs and continue to report progress, just like in Fig. 3.3. Then reduce jobs are started. In some cases, Hadoop can start to reduce jobs before all map jobs finish.

```
[hadoop@ip-172-31-47-40 ~]$ hadoop jar ~/contrib/streaming/hadoop-streaming.jar \
>    -mapper email_count_mapper.py \
>    -reducer email_count_reducer.py \
>    -input /data/hadoop.txt \
>    -output /data/output_hadoop_1 \
>    -file ~/email_count_mapper.py \
>    -file ~/email_count_reducer.py
14/09/02 10:40:11 WARN conf.Configuration: DEPRECATED: hadoop-site.xml found in the classpath. Usage of hadoop-site.xml is deprecated.
erties of core-default.xml, mapred-default.xml and hdfs-default.xml respectively
14/09/02 10:40:12 WARN streaming.StreamJob: -file option is deprecated, please use generic option -files instead.
packageJobJar: [/home/hadoop/email_count_mapper.py, /home/hadoop/email_count_reducer.py, /mnt/var/lib/hadoop/tmp/hadoop-unjar2903234617
14/09/02 10:40:16 INFO client.RMProxy: Connecting to ResourceManager at /172.31.47.40:9022
14/09/02 10:40:16 INFO client.RMProxy: Connecting to ResourceManager at /172.31.47.40:9022
14/09/02 10:40:18 INFO lzo.GPLNativeCodeLoader: Loaded native gpl library from the embedded binaries
14/09/02 10:40:18 INFO lzo.LzoCodec: Successfully loaded & initialized native-lzo library [hadoop-lzo rev 77cfa96225d62546008ca339b7c20
14/09/02 10:40:18 INFO mapred.FileInputFormat: Total input paths to process : 1
14/09/02 10:40:18 INFO mapreduce.JobSubmitter: number of splits:9
14/09/02 10:40:19 INFO mapreduce.JobSubmitter: Submitting tokens for job: job_1409651634781_0001
14/09/02 10:40:19 INFO impl.YarnClientImpl: Submitted application application_1409651634781_0001
14/09/02 10:40:19 INFO mapreduce.Job: The url to track the job: http://172.31.47.40:9046/proxy/application_1409651634781_0001/
14/09/02 10:40:19 INFO mapreduce.Job: Running job: job_1409651634781_0001
14/09/02 10:40:37 INFO mapreduce.Job: Job job_1409651634781_0001 running in uber mode : false
14/09/02 10:40:37 INFO mapreduce.Job:  map 0% reduce 0%
14/09/02 10:40:53 INFO mapreduce.Job:  map 6% reduce 0%
14/09/02 10:40:55 INFO mapreduce.Job:  map 11% reduce 0%
14/09/02 10:41:05 INFO mapreduce.Job:  map 15% reduce 0%
14/09/02 10:41:08 INFO mapreduce.Job:  map 18% reduce 0%
14/09/02 10:41:11 INFO mapreduce.Job:  map 26% reduce 0%
14/09/02 10:41:14 INFO mapreduce.Job:  map 29% reduce 0%
14/09/02 10:41:15 INFO mapreduce.Job:  map 40% reduce 0%
14/09/02 10:41:16 INFO mapreduce.Job:  map 44% reduce 0%
14/09/02 10:41:31 INFO mapreduce.Job:  map 50% reduce 0%
14/09/02 10:41:32 INFO mapreduce.Job:  map 50% reduce 5%
14/09/02 10:41:33 INFO mapreduce.Job:  map 56% reduce 5%
14/09/02 10:41:35 INFO mapreduce.Job:  map 56% reduce 6%
14/09/02 10:41:48 INFO mapreduce.Job:  map 61% reduce 6%
14/09/02 10:41:50 INFO mapreduce.Job:  map 67% reduce 6%
14/09/02 10:41:54 INFO mapreduce.Job:  map 67% reduce 7%
14/09/02 10:42:06 INFO mapreduce.Job:  map 72% reduce 7%
14/09/02 10:42:08 INFO mapreduce.Job:  map 78% reduce 7%
14/09/02 10:12:00 INFO mapreduce.Job:  map 78% reduce 9%
14/09/02 10:42:23 INFO mapreduce.Job:  map 84% reduce 9%
14/09/02 10:42:25 INFO mapreduce.Job:  map 89% reduce 9%
14/09/02 10:42:27 INFO mapreduce.Job:  map 89% reduce 10%
```

Fig. 3.3 MapReduce program running

Listing 3.6 Testing code without any Hadoop installation

```
1 cat hadoop_1m.txt | ./email_count_mapper.py | sort -k1,1 |
      ./email_count_reducer.py
2 python count_sum.py -r inline hadoop_1m.txt
```

Before you run the code on the cluster, you can test its basic functionality on a local machine. In Listing 3.6, I show you a simple common trick to simulate Hadoop workflow without Hadoop. The only difference is that the processing will not be distributed. This means that you might not be able to discover all potential problems with your code. At this point, just take it for granted that it works, especially the meaning of the sort operation. You will soon understand it better. Line 1 is an approach you would you if you plan to use Hadoop Streaming, and line 2 is an equivalent for MRJob

Listing 3.7 Sample content from the results of the MapReduce job

```
1 126.com 62
2 Colorado.edu 3
3 Honeywell.com 28
4 MEDecision.com 18
5 Zawodny.com 10
6 adobe.com 70
```

```
 7  amazon.com 35
 8  anyware-tech.com 26
 9  apache.org 144144
10  apple.com 47
11  biz360.com 68
12  blue.lu 19
13  carnival.com 10
14  ccri.com 22
15  clickable.com 60
16  collab.net 35
17  cse.unl.edu 379
```

After the code finishes running on the cluster, you can have a look at the results. Just run line 04 from Listing 3.4 and 3.5. You will see a list of domains with the frequency of appearance in the dataset. Some sample values can be found in Listing 3.7. The results are not perfect you can see some domains are obviously incorrect, especially if you inspect all the results. Dirty data is a common issue in data-intensive applications. I challenge you to improve on the algorithm I provide in the exercises at the end of this chapter.

3.4   Analyzing Structured Data

This is already the last section of this chapter. The ideas presented here will be fairly simple compared to earlier sections, especially for those of you who are familiar with SQL. I will show you how to perform a simple structured data analysis using Hive. Hive is a Hadoop tool that allows you to write simple SQL-like queries. It depends on data having a fixed structure, typically in a form of a CSV file. There are many datasets that come in a CSV format. It is also common to process unstructured data using MapReduce to produce structure files like CSV.

Figure 3.4 shows a simple overview of Hive workflow. Hive takes your query written in HiveQL and translates it to a series of MapReduce jobs. These jobs are executed automatically in a way not much different from what you have seen in the previous section. Finally, depending on the query, results are promptly returned to Hive.

Listing 3.8   Analyzing structured data with Hive

```
1  #start hive from Sandbox or cluster
2  hive
3
4  #create table and import data
5  CREATE EXTERNAL TABLE hadoopmail (list STRING, date1 STRING, date2
        STRING, email STRING, topic STRING)
6  ROW FORMAT DELIMITED FIELDS TERMINATED BY ',' LINES TERMINATED BY
        '\n'
7  STORED AS TEXTFILE;
8
```

```
 9 LOAD DATA INPATH '/dis_materials/hadoop.csv' OVERWRITE INTO TABLE
       hadoopmail;
10
11 #run a simple data query
12 SELECT substr(email, locate('@', email)+1), count(substr(email,
       locate('@', email)+1)) FROM hadoopmail GROUP BY substr(email,
       locate('@', email)+1);
```

It is easy to run your first job in Hive using the example CSV file that comes with this book. You can see a short snippet of it in Fig. 3.5. Example queries are in Listing 3.8. In line 02, you start hive with hive command from cluster prompt. First, you have to create a new table, using the query from lines 05–07. When the table is created, you load data in the table using LOAD statement from line 09. Remember that file location refers to HDFS. Finally, you can run a simple query in line 12. This query produces effectively the same results as the MapReduce job from the previous section.

You can see all the Hive queries from Listing 3.8 running in Fig. 3.6. Most interesting to notice is the familiar information about map and reduce job percent execution. Hive also provides you with a reference to the actual MapReduce job for monitoring or maintenance purposes.

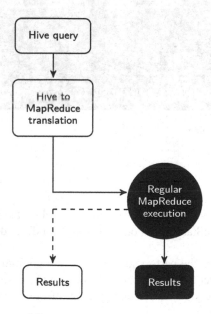

Fig. 3.4 Overview of Hive workflow

	A	B	C	D
1	general.hadoop.apache.org	Fri, 1 Oct 2010 12:06:54 -0300	marcoscba@gmail.com	Is it possible write directly to datanode's?
2	general.hadoop.apache.org	Fri, 1 Oct 2010 21:27:22 -0700	rmalviya@apple.com	Total Space Available on Hadoop Cluster Or Hadoop version of df"."
3	general.hadoop.apache.org	Sat, 2 Oct 2010 16:13:18 +1000	Glenn.Gore@melbourneit.com.au	RE: Total Space Available on Hadoop Cluster Or Hadoop version of df"."
4	general.hadoop.apache.org	Sat, 2 Oct 2010 08:12:48 -0400	palmercliff@gmail.com	Re: Is it possible write directly to datanode's?
5	general.hadoop.apache.org	Sat, 2 Oct 2010 10:39:09 -0400	marcoscba@gmail.com	Re: Total Space Available on Hadoop Cluster Or Hadoop version of df"."
6	general.hadoop.apache.org	Sat, 2 Oct 2010 09:52:42 -0700	rmalviya@apple.com	Re: Total Space Available on Hadoop Cluster Or Hadoop version of df"."
7	general.hadoop.apache.org	Sat, 2 Oct 2010 09:54:59 -0700	rmalviya@apple.com	Re: Total Space Available on Hadoop Cluster Or Hadoop version of df"."
8	general.hadoop.apache.org	Sat, 2 Oct 2010 11:57:50 -0700	ryanobjc@gmail.com	Re: Is it possible write directly to datanode's?
9	general.hadoop.apache.org	Sat, 2 Oct 2010 17:16:08 -0400	marcoscba@gmail.com	Re: Is it possible write directly to datanode's?
10	general.hadoop.apache.org	Sun, 3 Oct 2010 04:32:52 +0000	jgray@facebook.com	RE: Total Space Available on Hadoop Cluster Or Hadoop version of
11	general.hadoop.apache.org	Sun, 3 Oct 2010 10:58:13 -0700	rmalviya@apple.com	Re: Total Space Available on Hadoop Cluster Or Hadoop version of df"."
12	general.hadoop.apache.org	Mon, 4 Oct 2010 09:48:54 -0300	From general-return-2144-apmail-hadoop-general-archive=hadoop.apache.org@hadoop.apache.org Mon Oct 04 12:49:27 2010	
13	Return-Path: <general-return-2 Re: Total Space Available on Hadoop Cluster Or Hadoop version of df"."			
14	general.hadoop.apache.org	Wed, 6 Oct 2010 13:17:17 -0700	jdcryans@apache.org	[ANN] HBase-0.89.20100924, our third 'developer release', is now
15	general.hadoop.apache.org	Fri, 8 Oct 2010 15:01:09 -0500	davidlary@gmail.com	DFS error
16	general.hadoop.apache.org	Sat, 9 Oct 2010 22:14:46 +1100	eltonsky9404@gmail.com	Re: DFS error
17	general.hadoop.apache.org	Sat, 9 Oct 2010 22:01:24 -0500	davidlary@gmail.com	Re: DFS error
18	general.hadoop.apache.org	Mon, 11 Oct 2010 12:08:21 +0530	ssssssenator@gmail.com	Task tracker failed to start
19	general.hadoop.apache.org	Mon, 11 Oct 2010 19:52:08 +0800	hadoop.li@gmail.com	Re: DFS error
20	general.hadoop.apache.org	Mon, 11 Oct 2010 08:20:56 -0400	magawake@gmail.com	newbie setup question
21	general.hadoop.apache.org	Tue, 12 Oct 2010 06:11:02 -0700	hammer@cloudera.com	Re: newbie setup question
22	general.hadoop.apache.org	Tue, 12 Oct 2010 21:00:45 -0700	From general-return-2153-apmail-hadoop-general-archive=hadoop.apache.org@hadoop.apache.org Wed Oct 13 04:01:26 2010	
23	Return-Path: <general-return-2 eral-return-2153-apmail-hadoop-general-archive=hadoop.apache.org@hadoop.apache.org Wed Oct 13 04:01:26 2010			
24	general.hadoop.apache.org	Wed, 13 Oct 2010 09:32:00 +0530	paliwalashish@gmail.com	Re: Mailing list for posting a job opportunity
25	general.hadoop.apache.org	Wed, 13 Oct 2010 14:57:25 +0200	a.reiter@web.de	Architecture
26	general.hadoop.apache.org	Wed, 13 Oct 2010 15:09:33 +0200	timrobertson100@gmail.com	Re: Architecture
27	general.hadoop.apache.org	Wed, 13 Oct 2010 09:02:40 -0500	jon.creasy@announcemedia.com	Re: Architecture
28	general.hadoop.apache.org	Wed, 13 Oct 2010 16:36:47 +0200	lars.francke@gmail.com	Re: Architecture
29	general.hadoop.apache.org	Wed, 13 Oct 2010 07:46:01 -0700	omalley@apache.org	Re: Architecture
30	general.hadoop.apache.org	Wed, 13 Oct 2010 17:54:51 +0300	dslkar@gmail.com	Re: Architecture

Fig. 3.5 Sample content from hadoop.csv file

```
hive> CREATE EXTERNAL TABLE hadoopmail(list STRING, date1 STRING, date2 STRING, email STRING, topic STRING)
    > ROW FORMAT DELIMITED FIELDS TERMINATED BY ',' LINES TERMINATED BY '\n'
    > STORED AS TEXTFILE;
OK
Time taken: 0.591 seconds
hive> LOAD DATA INPATH '/data/hadoop.csv' OVERWRITE INTO TABLE hadoopmail;
Loading data to table default.hadoopmail
rmr: DEPRECATED: Please use 'rm -r' instead.
Deleted /mnt/hive_0110/warehouse/hadoopmail
Table default.hadoopmail stats: [num_partitions: 0, num_files: 1, num_rows: 0, total_size: 560404, raw_data_size: 0]
OK
Time taken: 1.284 seconds
hive> SELECT substr(email, locate('@', email)+1), count(substr(email, locate('@', email)+1)) FROM hadoopmail GROUP BY substr(email, locate('@', email)+1);
Total MapReduce jobs = 1
Launching Job 1 out of 1
Number of reduce tasks not specified. Estimated from input data size: 1
In order to change the average load for a reducer (in bytes):
  set hive.exec.reducers.bytes.per.reducer=<number>
In order to limit the maximum number of reducers:
  set hive.exec.reducers.max=<number>
In order to set a constant number of reducers:
  set mapred.reduce.tasks=<number>
Starting Job = job_1409651634781_0002, Tracking URL = http://172.31.47.40:9046/proxy/application_1409651634781_0002/
Kill Command = /home/hadoop/bin/hadoop job  -kill job_1409651634781_0002
Hadoop job information for Stage-1: number of mappers: 1; number of reducers: 1
2014-09-02 10:45:19,136 Stage-1 map = 0%,  reduce = 0%
2014-09-02 10:45:35,214 Stage-1 map = 100%,  reduce = 0%, Cumulative CPU 3.07 sec
```

Fig. 3.6 Hive program running

3.5 Exercises

Exercise 3.1 Evaluate and explain use cases for when to use AWS (or another public Cloud), OpenStack (or another private Cloud), and when to use a local Sandbox.

Exercise 3.2 Compare the performance of different configurations of OpenStack or AWS (e.g., different amount of nodes), running the same code.

Exercise 3.3 Improve your MapReduce program from Exercise 3 to account for any dirty data.

Exercise 3.4 Create MapReduce program that processes unstructured emails to CSV format. For an easy version, process data from hadoop.txt file. For an advanced version, process directly from AWS datasets. If you find this exercise too difficult right now, come back to it after completing Chap. 7.

Chapter 4
Functional Abstraction

The following sections deal with functional abstraction, which forms the foundation for data-intensive systems. Functional programming concepts help you to better understand the motivation and functioning of *data-intensive systems*. I explain the basics of functional programming in Sect. 4.1, and use these basic skills to form a functional abstraction for data processing in Sect. 4.2. In Sect. 4.3, I showcase how functional abstraction helps in analyzing algorithms complexity.

It is possible to use and, to some extent, understand data-intensive systems without the knowledge presented in this chapter. This fairly simple overview, however, demonstrates the powerful simplicity of the subject, resulting in deeper comprehension to help with further practical and theoretical studies.

4.1 Functional Programming Overview

We will not learn functional programming in this book. We do, though, need a short introduction to understand the functional abstraction that is used for data processing, particularly in data-intensive systems. If you are familiar with functional programming, feel free to skip this section.

Functional programming and related functional languages are typically seen in contrast to imperative programming, which you are most familiar with from languages like C or Python. You might have also read about declarative programming, the most common example of which being SQL.

Imperative languages focus on change of state or, in simple terms, variables that are shared between various parts of the program. Procedural and object-oriented approaches add an extra layer of both complexity and usability to imperative approach, but they do not change the basic nature of it. Functional languages focus on the evaluation of functions and passing values between them. Historically, functional languages were mostly an academic focus, but they have been gaining noticeable pop-

T. Wiktorski, *Data-intensive Systems*, Advanced Information and Knowledge Processing, https://doi.org/10.1007/978-3-030-04603-3_4

ularity with languages like Scala, Haskell, and Clojure. Many imperative languages nowadays feature some functional elements due to how well these elements abstract data processing routines.

Let me stress two key elements: functional programming, as it is aptly named, is based on functions. In this way, it avoids changes to the state of the program, and so also to the data, although the last bit might be hard to grasp right away.

Operations in an imperative program modify variables that are outside the scope of particular operation (see Fig. 4.1). Such variables are accessible to all operations. In principle, this is not a problem if all operations happen sequentially in a predefined order; however, such an assumption is only true for simple programs. The larger program or data gets, the greater the need for parallelism. Analysis of large datasets is currently almost exclusively done in a parallel fashion to increase the processing speed.

To process data in a parallel fashion, we need to understand what can and cannot be parallelized. The imperative code does not yield itself easily to such analysis (see Sect. 4.3 for further details). On the other hand, the functional code does. The functional code has strictly defined inputs and outputs and does not perform any side modifications of data (see Fig. 4.2). As a result, each part of the code becomes autonomous in the sense that it depends only on its inputs and outputs, which are

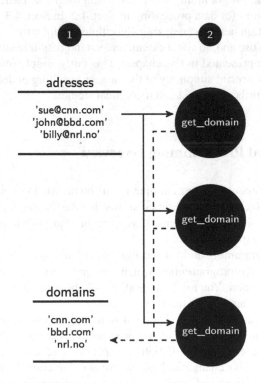

Fig. 4.1 Imperative program structure

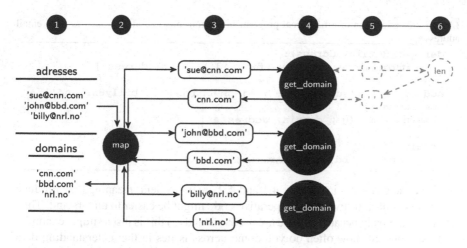

Fig. 4.2 Functional program structure

openly defined. This provides us with necessary information about the dependence of each function.

From a strictly functional programming perspective, though, this is not enough. Functional languages dictate that such program construction is the only possible way. This is in contrast to just using functional-inspired routines in imperative languages, e.g., Python. For now, it is enough to focus only on the general concepts, so we will not go into detail about how functional programs work, except for the basic routines.

As mentioned earlier, functional operations always create new data structures instead of modifying the existing ones; therefore, the order of operations does not matter unless they are explicitly chained. This ensures a lack of so-called "side effects," or simply modifying elements from other parts of the program parts might want to access. We will not go further with the explanation of this because it would be outside of the scope of this book. Be aware, though, that this particular feature will not be clearly visible in the examples below. The code is in Python, which is an imperative language and that we are only using in a functional way.

Listing 4.1 Execution of an imperative program in Python to extract domain names from email addresses

```
1 def get_domain(address):
2     return address[address.find('@')+1:len(address)]
3
4 addresses = ['sue@cnm.com', 'john@bbd.com', 'billy@nrl.no']
5 domains = []
6 for i in range(len(addresses)):
7     domains.append(get_domain(addresses[i]))
8
9 domains
10 ['cnm.com', 'bbd.com', 'nrl.no']
```

Listing 4.2 Execution of a functional program in Python to extract domain names from email addresses

```
1 def get_domain(address):
2     return address[address.find('@')+1:len(address)]
3
4 addresses = ['sue@cnm.com', 'john@bbd.com', 'billy@nrl.no']
5 domains = []
6 domains = map(get_domain, addresses)
7
8 domains
9 ['cnm.com', 'bbd.com', 'nrl.no']
```

The code examples in Listings 4.1 and 4.2 are equivalent, meaning they produce the same result. Intuitively, the imperative code might be easier to understand. This, however, is mostly because (1) you are used to it and (2) this is just a simple example. Just ask yourself: how often do you come across issues in the understanding data flow in the code, especially if code exceeds several pages? Functional code enforces a certain way of thinking about problems that help maintain code clarity.

These two examples introduce another feature of functional programming called high-order functions, or functions that take other functions as arguments. This is possible because functions are treated as first-class members of the program. Such a feature is now not exclusive to functional languages; you can implement it even in Java. It is not, however, a natural thing to do and is rarely used. The exception is Python, which now provides very good and simple support for such functionality. This capability is crucial for further data processing abstraction.

Some of the most typical high-order functions include *map*, *reduce* (or *fold* depending on language), and *filter*. Python 2 supports all of them in the built-in library. Python 3 supports map and filter in the built-in library, while reduce was moved to `functools`. All code was tested in Python 2 for clarity (you can also run it under Python 3 with minor changes).

Listing 4.3 Using lambda function in functional programs

```
1 addresses = ['sue@cnm.com', 'john@bbd.com', 'billy@nrl.no']
2 domains = map(lambda x: x[x.find('@')+1:len(x)], addresses)
3 domains
4 ['cnm.com', 'bbd.com', 'nrl.no']
```

Before we jump into the next section, let us also introduce something called *lambda*. You should simply think of lambda as an anonymous function, in that it is not bound to a name. We will use it in some examples to define simple functions on the fly. Listing 4.3 shows an example of using lambda applied to Listing 4.2.

Despite these benefits, there are also some downsides to functional programming. Functional programming will often take more memory than imperative and can be slower, especially comparing to C. The main reason for this is that function calls and immutable objects are expensive. Moreover, most programmers have only anecdotal familiarity with functional programming, so it might require additional time and effort, especially in the beginning, to understand and use functional concepts.

4.2 Functional Abstraction for Data Processing

In this section, we will review two major high-order functions that you will use throughout this book, though for a larger amount of data. These two functions are the map and reduce. I will also show examples of a third high-order function called filter. While you will not use filter function explicitly later, you will notice that the map function in the Hadoop framework has the combined functionality of traditional map and filter.

As you have seen examples with the map already in the previous section, we will now review it in detail, and do the same to reduce and filter functions. The map takes a list provided as a second argument and applies the processing function provided as the first argument to each element of the list. The map then outputs a list of the same length as the input list with each element transformed by the processing function.

Listing 4.4 Using map high-order function

```
1 addresses = ['sue@cnm.com', 'john@bbd.com', 'billy@nrl.no']
2 domains = map(lambda x: x[x.find('@')+1:len(x)], addresses)
3 domains
4 ['cnm.com', 'bbd.com', 'nrl.no']
```

Please take a look at Listing 4.4. In line 1, I define a list called `addresses` consisting of three elements. This will be our input list. In line 2, I then define a new variable `domains` that will be our output list, to `domains`. We assign the output of the map function. The map has two parameters separated by a comma; the first parameter is the processing function. Here I use lambda, which means we define an anonymous function. The function is `x[x.find('@')+1:len(x)]`, which is a simple substring selection. The function we just defined is applied by map to each element of `addresses` list, being sue@cnm.com, john@bbd.com, etc. Each value is then added to the domains list. The result is in line 4.

The filter takes a list provided as a second argument and applies the processing function provided as the first argument to each element of the list; it then outputs only those elements from the first list for which the processing function returned true. Elements themselves are untransformed.

Listing 4.5 Using filter high-order function

```
1 addresses = ['sue@cnm.com', 'john@bbd.com', 'billy@nrl.no']
2 com_addresses = filter(lambda x: True if x.find('.com')>-1
      else False, addresses)
3 com_addresses
4 ['cnm.com', 'bbd.com']
```

Please take a look at Listing 4.5. In line 1, I again define a list called `addresses` consisting of three elements, which will be our input list. In line 2, I then define a new variable `com_addresses` that will be our output list. To it, I assign the output of the filter function. The filter has two parameters separated by a comma; again, the first one is the processing function. I use lambda again, and the function is `True` `if x.find('.com')>-1 else False`. The output of this function is TRUE for any x that contains a .com domain and FALSE for any x that does not contain it. This function is then applied to each element from `addresses` list. Each value from the `addresses` list which returns TRUE is added to `com_addresses` list. The result is in line 4.

We see now that map outputs lists of exactly the same length as the input lists. Filter outputs lists of the same or smaller length as the input lists. The filter also does not perform any transformation on the data. You will see later that filter functionality can be included in the map functionality. We will not, therefore, be using filter directly, though we will be using its functionality.

The last function I want to talk about is reduce. Reduce takes a list provided as a second argument and applies the processing function, provided as the first argument, continually to pairs of elements. The first element of each pair is the result of combining the last pair; the second element is the next unprocessed element on the list. Let me explain this in an example:

Listing 4.6   Using reduce high-order function

```
1 addresses = ['sue@cnm.com', 'john@bbd.com', 'billy@nrl.no']
2 total_length = reduce(lambda x,y: x+len(y) if type(x) is int
      else len(x)+len(y), addresses)
3 total_length
4 35
```

Please take a look at Listing 4.6. In line 1, I again defined a list called `addresses` consisting of three elements; this will be our input list. In line 2, I then defined new variable `total_length` that will be our output variable (not a list), to `total_length` I assign the output of reduce function. Reduce has two parameters separated by a comma. First one is the processing function. Here, I use lambda, which means I define an anonymous function on the spot. The function is `x+len(y)`. This function is then applied continually to pairs of elements in the list. The result is in line 4.

To better understand how the reduce works, let us look at the intermediate steps in our simple example:

`Step 1. x='sue@cnm.com', y='john@bbd.com', result=23`
`Step 2. x=23, y='billy@nrl.no', result=35`

This is our final result.

Listing 4.7   Another example of reduce high-order function

```
1 longest_domain = reduce(lambda x,y: x if (len(x)>len(y)) else
      y, domains)
2 longest_domain
3 'bbd.com'
```

In Listing 4.7, you can find another simple example of a reduce function. I will leave the analysis to you.

These three functions we analyzed here are simple, but at the same time powerful. They have clearly defined execution semantics, meaning we can easily know how they will be executed regardless of what the actual function being applied to the list elements is.

This creates a useful basis for designing parallel algorithms for large datasets. We will take a look at how this abstraction support parallelization in Sect. 4.3.

4.3 Functional Abstraction and Parallelism

In this section, I will explain how basic concepts of functional programing can help in writing parallel programs.

Imperative programs are inherently sequential; statements are executed one after another. It is possible to parallelize imperative programs through the use of, for example, threads. It is, however, a difficult process. Imperative programs have side effects, which means they can modify variables other than their outputs. These variables can be accessed by other concurrent processes, which leads to "race conditions." Another difficulty is that parallelization has to be approached explicitly and often individually for each functionality you want to implement. These problems can be solved using functional abstraction and high-order functions.

I would like you to understand better what race conditions are, which sometimes are also called *race hazards*. If the result your code produces is dependent on other operations (in particular on their timing), which are outside your control, you created a race condition. Let me take you through several examples.

Listing 4.8 Two parallel processes using the same variable

```
1 address = 'sue@cnm.com'         ...
2 domain = get_domain(address)    ...
3 ...                             address = 'john@bbd.com'
4 ...                             length = len(address)
5 domain                          length
6 'cnm.com'                       12
```

Listing 4.9 Example 1 of race condition

```
1 address = 'sue@cnm.com'         ...
2 ...                             address = 'john@bbd.com'
3 domain = get_domain(address)    ...
4 ...                             length = len(address)
5 domain                          length
6 'bbd.com'                       12
```

Listing 4.10 Example 2 of race condition

```
1  ...                              address = 'john@bbd.com'
2  address = 'sue@cnm.com'          ...
3  ...                              length = len(address)
4  domain = get_domain(address)     ...
5  domain                           length
6  'cnm.com'                        11
```

Listing 4.8 shows two processes that attempt to access the same variable at two different moments of program execution. The expected result after executing the program is `domain='cnn.com'` and `length=12`. If these two processes run in parallel, they have a race condition for variable tmp. Since programs can run at different speeds, it is impossible to guarantee the correct execution of either program. Listings 4.9 and 4.10 illustrate two possible problem scenarios.

To avoid race conditions, the programmer must use special synchronization and isolation mechanisms that prevent processes from interfering with one another. Not only does this require additional effort to implement, it is also easy to miss race conditions. As a result, creating and debugging parallel programs written using imperative paradigm is difficult and time-consuming, especially since potential problems are almost impossible to reproduce.

Functional abstraction eliminates these problems by eliminating side effects. The only modified variable is the declared output. Moreover, high-order functions like map, filter, and reduce have predefined parallel behavior. Program parallelization is then achieved automatically by expressing algorithms through combinations of such functions. It has been long thought that such an approach is too limiting, but recent years have shown that a surprising amount of algorithms can be coded this way and that the forced simplicity sparked creativeness in programmers rather than deterred it.

Map function—as well as filter function—is inherently parallel. This means that, per definition, its execution can be performed on chunks of the input list without any influence on results. Take a look at the Fig. 4.3. We divide an input list of emails into chunks, each chunk processed by the same function, e.g., extracting domain name (check Listing 4.3 for reference).

Reduce function is on the other hand inherently sequential. This means that, per definition, its execution has to be performed on the whole list step by step; otherwise, results might be influenced. Take a look at Fig. 4.4.

It is important to note that reduce can be parallelized in two general cases: (1) for associative operators (sum is an associative operator), and (2) if the list can be divided into sub-lists by distinctive values of the key. For now, just keep this in mind. We will go into detail later in the course.

If your code makes use of the functional concepts, this means there is no shared state and data is passed around as inputs and outputs of functions, which you can then use to express dependencies between functions as a Direct Acyclic Graph (DAG). Based on that graph, you can determine that any two or more functions can run

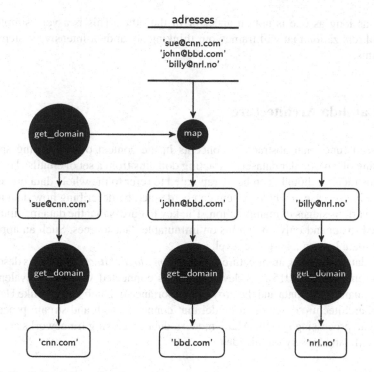

Fig. 4.3 Parallelism of map function

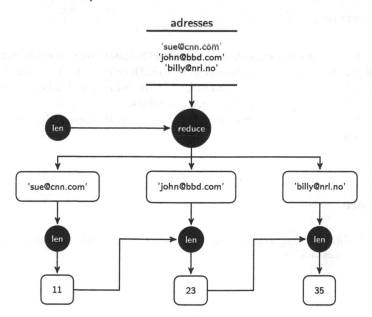

Fig. 4.4 Lack of parallelism of reduce function

parallel as long as one is not an ancestor of the other. This is a very simple and powerful realization that will frame your thinking about data-intensive systems and algorithms.

4.4 Lambda Architecture

I discussed functional abstraction concepts in the context of performing specific processing of a particular dataset, extracting domains from a set of emails. The functional abstraction, though, can be extrapolated to refer to the whole data processing systems. In this case, a whole system can be defined by describing how data moves through various stages of manipulation. The key feature is that the data manipulation is limited to append-only operations on immutable data sources. Such an approach prevents data loss on a systemic level.

Such data processing architecture is called *lambda architecture* and was described by Marz and Warren (2015). Its development is connected with the prevalence of massive quantities of data and the growing importance of batch systems like Hadoop. Lambda architecture describes a model that connects batch and stream processing in one data processing system. Many modern data processing frameworks are based both strictly and loosely on this idea.

4.5 Exercises

Exercise 4.1   Think of a small dataset. Construct at least two examples of each map and reduce operations that you could apply to it. Draw a graph. Write pseudocode.

Exercise 4.2   Based on the previous exercise, try to link map and reduce operations. Think of potential parallelism and race conditions.

Exercise 4.3   Investigate lambda architecture: what are the typical elements, where are they used?

Reference

Marz N, Warren J (2015) Big data: principles and best practices of scalable realtime data systems. Manning Publications Co

Chapter 5
Introduction to MapReduce

This chapter introduces you to MapReduce programming. You will see how functional abstraction lead to real-life implementation. There are two key technical solutions that enable the use of map and reduce functions in practice for parallel processing of big data. First of all, a distributed file system, like Hadoop Distributed File System (HDFS), which ensures delivery of unique subsets of the whole dataset to each map instance. These subsets are called *blocks*. For the duration of this chapter, I ask you to assume it just works. You will learn more about it in Chap. 6.

The second technical solution is structuring of data processing with key–value pairs. Keys allow the MapReduce framework, like Hadoop, to control the data flow through the whole pipeline from HDFS, through map, combine, shuffle, reduce, and HDFS again. Keys are also a way for you to get the most of the framework. Map and reduce together with the key–value pairs will be a way for you to express your algorithms through working with this book.

Many problems can be expressed in terms of map and reduce operations. Most of ETL tasks are a natural fit, also analytic workloads and some machine learning algorithms can be adapted. Google was the first to apply MapReduce approach to large-scale data processing with its work on processing search index. Hadoop is currently the main open-source implementation of this concept. Many commercial versions exist, which usually are just extensions of Hadoop. The benefits they provide are typically performance improvements, integration with other systems, and tech support. Some of the commercial versions are available for free for nonprofit use.

In addition to Hadoop, Spark is one of the most popular frameworks. However, its implementation reflects the current trend of growing importance of in-memory processing. This improves the performance of many algorithms, but can also lead to scalability limitations, depending on the hardware architecture and algorithm's implementation. Both Hadoop and Spark can be used in parallel, complementing each other's functionality.

© The Author(s), under exclusive license to Springer Nature Switzerland AG 2019   41
T. Wiktorski, *Data-intensive Systems*, Advanced Information and Knowledge Processing,
https://doi.org/10.1007/978-3-030-04603-3_5

5.1   Reference Code

Listing 5.1 Running Hadoop job using Hadoop Streaming

```
1 hadoop jar ~/contrib/streaming/hadoop-streaming.jar \
2   -mapper email_count_mapper.py \
3   -reducer email_count_reducer.py \
4   -combiner email_count_reducer.py \
5   -input /data/hadoop.txt \
6   -output /data/output_hadoop_1 \
7   -file ~/email_count_mapper.py \
8   -file ~/email_count_reducer.py
9
10 hadoop fs -text /data/output_hadoop_1/part* | less
```

Listing 5.2 Running Hadoop job using MRJob

```
1 python count_sum.py -r hadoop hdfs:///dis/hadoop.txt
    --output-dir hdfs:///dis/output1
2
3 hadoop fs -text /dis/output1/part* | less
```

Listing 5.3 Counting mapper for Hadoop Streaming

```
1 import sys
2
3 for line in sys.stdin:
4     line = line.strip()
5     if line.find("From:") == 0:
6         email_domain = line[line.find("@")+1:line.find(">")]
7         if len(email_domain) == 0:
8             email_domain == "empty"
9         print '%s\t%s' % (email_domain, 1)
```

Listing 5.4 Counting reducer for Hadoop Streaming

```
1 from operator import itemgetter
2 import sys
3
4 current_email_domain = None
5 current_count = 0
6 email_domain = None
7
8 for line in sys.stdin:
9     line = line.strip()
10    try:
11        email_domain, count = line.split('\t', 1)
12    except:
13        print "error: ", line
14        continue
15
```

```
16    try:
17        count = int(count)
18    except ValueError:
19        continue
20
21    if current_email_domain == email_domain:
22        current_count += count
23    else:
24        if current_email_domain:
25            print '%s\t%s' % (current_email_domain,
                    current_count)
26        current_count = count
27        current_email_domain = email_domain
28
29 if current_email_domain == email_domain:
30    print '%s\t%s' % (current_email_domain, current_count)
```

Listing 5.5 Counting with MRJob

```
1 from mrjob.job import MRJob
2
3 class MRCountSum(MRJob):
4
5    def mapper(self, _, line):
6        line = line.strip()
7        if line.find("From:") == 0:
8            email_domain = line[line.find("@")+1:line.find(">")]
9            if len(email_domain) == 0:
10                email_domain == "empty"
11            yield email_domain, 1
12
13    def combiner(self, key, values):
14        yield key, sum(values)
15
16    def reducer(self, key, values):
17        yield key, sum(values)
18
19
20 if __name__ == '__main__':
21    MRCountSum.run()
```

Through this chapter, I will explain in detail how the commands in Listings 5.1 and 5.2, and related code in Listings 5.3, 5.4, and 5.5, are in fact executed by the Hadoop MapReduce. While I explain each part, notice how these complex processes require just a few lines of code. The framework makes many assumptions about how your code is written. You have to conform with the expected functional abstraction and key–value pair's structure. If you do, write MR programs will be simple and your programs will be efficient and scalable.

5.2 Map Phase

The map is the first phase which is going to be executed by MR framework. At step 1 in Fig. 5.1, data is being read from HDFS file or directory. Line 6 in Listing 5.1 and Line 1 in Listing 5.2 photo the data source in HDFS. Data in HDFS is pre-divided in blocks, more on this in Chap. 6. In our case, there is one file, which contains a set of emails. Each block will then contain a subset of these emails.

At the moment the blocks are transferred to individual map task at step 2, they become known as splits. Each map task receives a split to process. How the data is split in part is controlled by the MR framework. You can modify this process by writing a custom splitter. It is uncommon to do so, however, and in most cases, you can rely on the standard behavior.

Data supplied to map tasks does not necessarily have key–value structure. However, the framework will introduce such formal structure regardless. For example in case of simple text files, the byte offset of text line will be used as the key and the text as the value. If the input conforms to some key–value scheme, a special input format can be used. By default, Hadoop expects text files and processes them line by line. This is exactly our case.

A separate map task is executed for each split at step 3, but they all run the same map function which you provide in Line 3 of the reference code. Map tasks can be executed on the same or several separate machines in the cluster. Allocation of map tasks is decided by the scheduler. It depends on many factors, but to the largest extent data location and amount of other tasks being concurrently executed. The amount of map tasks is decided based on the size of input data and parameters of the processing.

In our code for map function, we ignore the keys on the input instead, and we group the lines in the split together and processes them as the whole. At step 4,

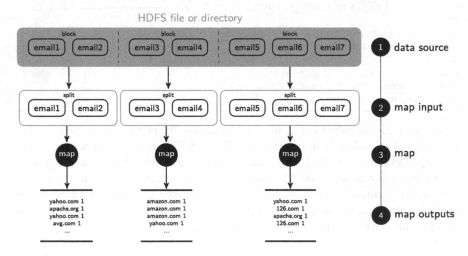

Fig. 5.1 Example of map phase

the output is produced. Here, we make use of key–value structure, and we choose the email address for the key and output *1* as the value. This is a standard way to implement counting of any sort in MR. A key–value part with value *1* is output for each element to be counted, which key being this element. These results are later summed up together in reduce phase. Map output is stored in a local file system of the node running each map. Map outputs do not have to be persisted as they are just an input to reduce. Therefore, HDFS is not used, as its write performance is worse than that of the local file system.

Sometimes, it is possible to introduce optimizations in map phase that would reduce the amount of data output to reduce. In our case, the wisest thing would be to try to locally sum the ones for each key. Depending on the format of data and algorithm used for map function, it is not always possible to introduce such optimization directly in the map phase. It might, for instance, lead to an increased complexity of the input format or convoluted implementation of map function. It is, nevertheless, highly desirable to perform this optimization. This is why it is common to perform it in a separate step called a *combiner*.

5.3 Combine Phase

The goal of the combine phase is to improve the execution performance of MR program. This is achieved by reducing the amount of data transferred to reduce phase. Because the network tends to be the slowest component, the impact of combining can be significant. Combining is sometimes called map-side reducing, but combine function does not have to be the same as reduce. It is, though, correct to think of it as some form of reduce from the perspective of functional abstraction.

Combine phase receives the input directly from map phase, at step 1 in Fig. 5.2. Because data is not transferred, this operation does not influence the performance as sending data to reduce would. Key–value structure of map output is maintained as combine input.

Combine function is executed for each map output independently. In our case, combine function, in Line 5 of reference code, is the same as reduce function. Finally, combine produces the output at step 3. This output is stored in a local file system, same as in case of map phase.

The most important restriction to keep in mind is that combine must maintain the key–value structure introduced during map phase, both as its input and output structure. While the input part is quite obvious, let us consider why the output structure also has to be the same.

When executing your job that contains combiner, you are in fact not guaranteed that the combine part will be executed. It is seen only as an optimization element and not as a necessary processing step. This approach is consistent with the origin of combine phase. In some older versions of Hadoop, it was not possible to use a non-Java combiner in streaming or it had to be the same as the reducer.

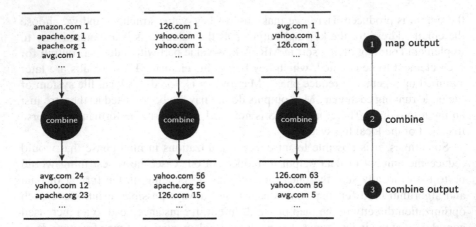

Fig. 5.2 Example of combine phase

It will differ from case to case how much performance improvement through data reduction we achieve. In this toy example, the hypothetical data reduction rate is over 75x. In practice, it might be difficult to achieve such numbers, but the reduction of 10x or more is not uncommon. Shuffling 1/10 of data to reduce phase will have a profound impact on the performance of the whole job.

5.4   Shuffle Phase

Shuffle is probably the most complex of all phases. At the same time, it is totally invisible to you and you do not have to program anything for it, except for maintaining a proper key–value structure of your data and related map and reduce functions. This is why you do not see it reflected in the reference code in Listings 5.1 and 5.2.

The purpose of the shuffle phase is to transfer data from a map (or combine) to reduce phase. Such a process is straightforward if there is only one reducer. In most cases, though, we would rather have multiple reducers to reduce the time this phase takes. This is the case we will consider further in this section.

As you might remember, the reduce operation is inherently sequential. However, it is possible to parallelize it if the operations you are using are associative. Sometimes, non-associative operations can be made such with just small modifications. For instance, arithmetic mean is not associative, but it can be turned into weighted arithmetic mean which is.

The shuffle phase starts with the output from a map or combine phase at step 1 in Fig. 5.3. Data arrives structured according to the keys defined in the earlier phase. At step 2, data is partitioned based on the key. The exact partitioning scheme is decided based on the key structure and amount of reducers. Data is also sorted within each partition at step 3 and it is ready to be shuffled to the reducers at step 4. Shuffle at step 5 is not a single function like map or sort, but a whole process.

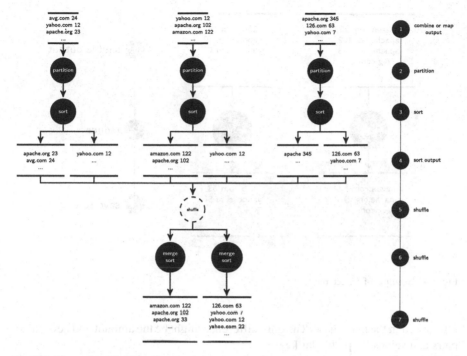

Fig. 5.3 Example of shuffle phase

Data after arriving on the reduce side are first merged and sorted again at step 6. Merging connects data with the same key that came from different map nodes and produces an output at step 7, which is fed to the reduce function.

Shuffle phase guarantees that data arriving at the reduce phase are grouped by key and sorted. This is not a simple guarantee to make. Consider that data for each key might be spread across many nodes and map phase might take different time to finish on each of them.

There is far more to the shuffle phase than I presented here. You can find all the details and technicalities in Hadoop documentation. I believe that an overview of the level just presented should suffice further work with Hadoop.

5.5 Reduce Phase

The purpose of the reduce phase is to produce final output of the MR job to store in HDFS. Reduce is the last stage of an MR job, but in many cases, a whole algorithm will contain several consecutive MR jobs. This, however, does not influence in principle how a single job runs.

Figure 5.4 presents an example of a reduce phase. Merged and sorted date are fed to reduce phase, at step 1. These data still have the same key–value structure as

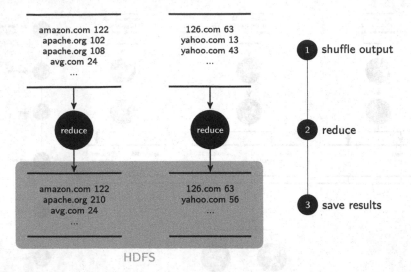

Fig. 5.4 Example of reduce phase

introduced in the map phase. The only difference might be the amount and key–value pairs and values for particular keys.

Reduce function is executed, at step 2, for each set of values corresponding to a different key. The reduce function is defined in Line 4 in the reference code. It is important to notice that different frameworks implement reduce slightly differently. Hadoop passes a list of values to the reduce together with key–value. This is a deviation from a pure functional definition of the reduce, as we see it implemented in Python or Spark. In these cases, only one value is passed at a time together with current aggregate.

The output from the reduce is saved in HDFS file or set of file in a directory, at step 3. This location is defined in Line 7 of the reference code. In principle, reduce can redefine the key structure, but it is not necessary in our case.

After the reduce phase, you will usually find several files in the output directory. This is due to several reducers running in parallel. It is enough to simply append these files together because their contents are independent due to reducers running on data grouped by key. You can see the results using the command in Line 12.

5.6 Embarrassingly Parallel Problems

When working further with Hadoop and data-intensive systems, you will often come across so-called embarrassingly parallel problems or algorithms. In general, it means that there is no communication between parallel tasks. In MapReduce terms, this simply means that the implementation only requires map phase and no reduce. In case of jobs requiring a reduce phase, but where very little data is transferred to it, the embarrassingly parallel label is also sometimes used.

Even for operations that seem clearly not embarrassingly parallel, there might be (1) a subset or (2) a part of the algorithm that can be easily parallelized to a greater general benefit. Join operations are typically quite complicated due to data being split between different nodes. However, a replicated join is embarrassingly parallel, an example of case (1). A replicated join connects two datasets, where one of them is limited in size to an extent that it can easily be replicated to each node in the cluster. In current practice, such datasets might reach several hundred megabytes, so they are not necessarily as small as you might originally think. Having one of the datasets replicated to each node makes it easy to implement join in map phase only.

An example of the case (2) is many machine learning algorithms, e.g., particle swarm optimization or random forests. The whole algorithm requires both map and reduce, and often several iterations of them. However, the training phase, which is often the most time-consuming, can be easily parallelized.

5.7 Running MapReduce Programs

You can choose between three general modes of executing a MapReduce program in Hadoop: standalone, pseudo-distributed, and distributed. Each of them has its unique purpose.

The *standalone* mode is a simple mode, which you can use to test your implementation. It runs on your local machine and executes everything in one Java process. HDFS is not being used, and all data is accessed from the local file system. The use of resources is limited, which helps with testing even on older machines. It is also easy to debug because only one java process is running.

In order to simulate some of the complexities of a real-world implementation, you can use *pseudo-distributed* mode. It runs Hadoop processes in separate java processes and uses HDFS. It allows you to discover many of the problems with dependencies between job, which could be left unnoticed under the standalone testing. You also run it on a single local machine, but it is more resource demanding.

Finally, *distributed* mode is the regular way to run Hadoop in a production environment. Hadoop process runs on separate physical machines connected in a network. Typically, we talk then about a *Hadoop cluster*. It is expected that the code deployed on a cluster is already debugged using two other modes to avoid destabilizing the cluster.

In recent years, it has become common to use pre-configured virtual machines and containers with Hadoop distributions for local tests. Such a solution has many advantages. You skip a difficult installation process, Hadoop environment is fully isolated from your local machine, and you can easily share your environment with other people. It is also the only easy way for Windows users to have a local installation of Hadoop at all. Hadoop can naively run on POSIX systems, like Linux or Mac OS, but not on Windows.

These are general modes that a Hadoop distribution uses. As long you use a ready Hadoop installation, e.g., in a container, the choice is made for you. MRJob offers, in addition, four running modes.

First two *inline* and *local* are designed for testing your code. *Inline* runs your job as a single process without using any Hadoop cluster. *Local* simulates multiprocess properties of a Hadoop environment, but still does not actually use any Hadoop installation. Doing first rounds of testing using these modes will greatly speed up your work and help in discovering general coding mistakes before you move to a Hadoop cluster.

The mode *Hadoop* uses available Hadoop installation and provides the most realistic environment to test your code. In our case, you are using the Hadoop installed in the Docker container, but it is possible to setup MRJob to access any other installation.

The last mode *emr* executes your code on an Elastic MapReduce cluster supplied by Amazon Web Services. By combining all modes, you can easily move from early-stage testing to full-scale execution without any additional effort.

5.8   Exercises

Exercise 5.1   In Chap. 3, we run the code without combiner, but its map produced the same result as map w/ combiner in this chapter, why?

Exercise 5.2   Think of cases when you can use your reducer as a combiner. When is it possible and when not? Consider about both practical and theoretical aspects.

Exercise 5.3   Modify the example from Chap. 3 to reflect map and combine logic introduced in this chapter. Measure data reduction rate for data sent to the reducer and change in performance of the whole algorithms. Use both a VM and a cluster. Compare and explain the differences.

Exercise 5.4   Search for more examples of embarrassingly parallel problems. Identify those you could apply to your projects.

Chapter 6
Hadoop Architecture

In this module, you will learn about the basic components of the Hadoop framework, how they work internally and how they are interconnected. I will primarily focus on the two main components: Hadoop Distributed File System (HDFS) and MapReduce (MR). These two components provide the basic Hadoop functionality that most other elements rely on. I will also shortly cover other components, mostly to provide you with a basis for further independent exploration.

Having dealt with these fundamentals, I will discuss data handling routines in Hadoop, in particular, write and read operations, followed by handling typical HDFS failures. From there, I will move to job flow, describing how data is circulated between MR components, considering data locality problem, and finishing with handling typical failures in MapReduce.

Understanding of Hadoop architecture forms the basis for any further work on applications and algorithms because the architecture has the major influence on the design of algorithms and their performance. Even though other data-intensive systems might have different architectures, the main conceptual elements will remain the same.

6.1 Architecture Overview

The basic Hadoop architecture consists of two main elements: Hadoop Distributed File System (HDFS) and MapReduce (MR) framework. HDFS is responsible for data storage and consists of name node and data nodes. MR is responsible for data processing and consists of job tracker and task trackers together with tasks. HDFS and MR processes run on the same machines and can communicate with each other directly. You can see a simple overview in Fig. 6.1.

Data nodes are responsible for the actual data storage. Name node coordinates the work of data nodes, spreading data among them, but does not store any data

© The Author(s), under exclusive license to Springer Nature Switzerland AG 2019   51
T. Wiktorski, *Data-intensive Systems*, Advanced Information and Knowledge Processing,
https://doi.org/10.1007/978-3-030-04603-3_6

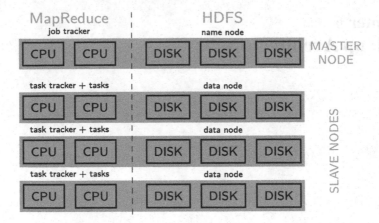

Fig. 6.1 Hadoop architecture overview

itself. Tasks that run on each node are controlled by a local task tracker. Central Job distributes all submitted tasks among task trackers. There is a clear parallel in the functioning of the data and task layers. Name node coordinates work of data nodes in an equivalent fashion as job tracker coordinates work of task trackers. Data nodes and task trackers with tasks are usually co-located. This reduces data bottlenecks in the system avoiding excessive network traffic. It allows for the practical implementation of the idea of sending computation to data. After the initial spreading of data across the cluster, job tracker then sends tasks to nodes that have the required data, by obtaining this information from name node.

There are many other elements in the Hadoop ecosystem. However, they are in principle all built on top of HDFS and MR. In Fig. 6.2, you can see an overview of the most common and important components representing typical functionalities. Please be aware that this is just a small subset and placement of components reflects their logical function, instead of software dependencies. For example, in storage layer, HBase is depicted in parallel with HDFS, which reflects its function, but does not reflect the fact that it is strictly dependent on HDFS for its operation.

Hue	Aginity	Jupyter	USER INTERACE
MapReduce Pig/Hive Mahout	Cascading Impala Solr	Spark SQL/Streaming MLlib	FRAMEWORKS and LIBRARIES
HDFS	HBase	Tachyon	STORAGE
Yarn	Corona	Mesos	RESOURCE MANAGEMENT

Fig. 6.2 Major components of Hadoop and data-intensive systems

All Hadoop clusters base on some form of resource management. YARN is a default resource manager for Apache Hadoop. Mesos is a lightweight alternative originating from the work on Spark. Each of them can support most all typical components and additional applications. What is most important is that you can run traditional MR on Mesos and Spark on YARN.

Storage is the next conceptual layer after resource management. HDFS is the basic distributed file system for all Hadoop installations. HBase is a distributed key–value database that excels at random and range lookups. It uses HDFS for basic storage but extends it with special indexing system and storage structures. Tachyon is an in-memory distributed storage system, a relatively new addition in comparison to HDFS and HBase, but an important one considering the growing importance of in-memory processing.

Various frameworks and libraries base on the storage layer. Traditional MR bases typically on HDFS, though being formally independent. Various systems were invented to simplify the creation of MR code, two most well known being Pig and Hive. Pig offers script-like language and Hive SQL-like queries; both are converted to regular MR job for execution. Mahout is probably the best-known library offering MR versions of typical machine learning algorithms.

Spark offers a similar set of functionalities as MR but adds support for in-memory processing. Basic Spark offers routines that could be compared to a combination of MR and Pig. In addition, a SQL-like language is available to write queries, which can be automatically executed as Spark jobs. MLlib holds a place similar to Mahout. Spark Streaming does not have an immediate counterpart in tradition MR framework.

I decided to single out four more frameworks that demonstrate a variety of innovation stemming from the central role of Hadoop for data-intensive processing. Cascading is a tool for designing and executing complex MapReduce data processing work flows. Impala is a distributed high-performance database with SQL query engine. Storm is a framework for operating on real-time unbounded streams of data. SOLR is integrated, scalable, and fault-tolerant search platform.

The last layer constitutes user interfaces. Historically, this has been the weakest part of the ecosystem. A state that to some extent persists to this day despite noticeable progress. Hue is a user interface giving access to all main components of the Hadoop stack: HDFS, Pig, Hive, etc. It also allows basic file operations, both within the HDFS, but also between HDFS and local file system. Jupyter is a universal notebook-like interface, originating from the work on IPython Notebooks. Recently, it has been gaining significant presence for data-intensive processing in connection with Spark.

Further, in this chapter, I focus only on details of the operation of MR and HDFS. Some other elements can be very important and you will be definitely using them in practice. However, they are not necessary in order to understand the basic concepts which are the practical extension of the theoretical idea of functional abstraction. I believe that the understanding of these basic concepts is the precondition for further exploration and application of data-intensive systems.

6.2   Data Handling

In this section, I will explain how the data is handled in HDFS. I will start by looking deeper at HDFS structure, then I present a general overview of write and read operations, and finally, I shortly discuss typical HDFS failures.

6.2.1 HDFS Architecture

As I mentioned earlier, the two main components of HDFS are name node and data nodes. In addition, we find more elements, which you can see in Fig. 6.3. In traditional HDFS architecture, based on Hadoop 1.0, we find also a secondary name node, which contrary to its name does not provide any backup functionality for the main name node. Its purpose is to update namespace with edit logs. Edit logs are applied to namespace image only at the restart of name node which, in typical circumstances, is seldom. This creates performance bottlenecks, which secondary name node aims to solve.

Because name node is the one element, we can never afford to lose it which is common to keep an additional copy of name nodes metadata files. In Hadoop 2.0, a new solution was introduced, called HDFS HA (High Availability). It is depicted on the right side of the figure. There is no secondary name node; instead, there are two name nodes, one active and one standby. Active name node instead of regular edit logs sends edits to shared logs which reside on journal nodes. It is necessary to provision at least three of them. These logs are then merged with namespace on the

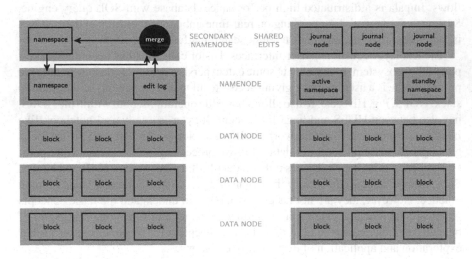

Fig. 6.3 HDFS building blocks

standby node, which this way replaces the functionality of secondary name node and also provides failover mechanism in case of main name node failure.

Each file when created is divided into blocks. When talking about MR usually, instead of blocks, we refer to parts of the file as splits. These terms are not theoretically equivalent, one referring to storage and the other to processing abstraction. However, in practice, they correspond to the same chunks of data. You will sometimes see these terms used interchangeably.

The blocks are distributed across the cluster; moreover, each block is typically kept in three copies. The amount of copies is called *replication factor*. The goal is to be able to read different parts of the file simultaneously on different cluster nodes to answer the data challenges we discussed at the beginning of this chapter. This also provides protection against disk failure, data node failure, and data loss in general. Name node maintains the information about the location of each block in the namespace.

There are four common ways of accessing data in HDFS:

- Command line—with POSIX-like syntax, most common for everyday data exploration and ingestion;
- Java API—most common during MR processing, also for languages other than Java, which typically go through Java API, but some might re-implementing access methods (good example is Spark);
- Hue—a GUI for everyday data exploration and ingestion;
- Flume—for ingesting large dataset on a regular basis.

6.2.2 Read Flow

The key element of HDFS design is its scalability. It means that, when amount of data grows, it is enough to add additional servers to the cluster to maintain performance. This is achieved by directly serving data from each data node, while the name node is used only to provide metadata, which is relatively small.

The read flow is presented in Fig. 6.4. In step 1, client opens the file and receives the location of the blocks from name node. In typical installations, data nodes and tasks are collocated physically and the scheduler will aim to allocate processing tasks to nodes that host the relevant data. If it is not possible, another location is chosen as close as possible to the client. The client reads data from the first block in step 2 and as soon as the block is finished, data node for the second block is found and clients start reading from it in step 3. In step 4, client continues to read data from subsequent blocks. The change of blocks and data nodes is hidden from the client, who observes it as one seamless operation. When all blocks are read, the process finishes and no additional action is required from either name node or data node. If reading of a particular block from a particular data node does not succeed, another data node will be used. Checksums for each block are verified, and corrupted blocks are reported to the name node.

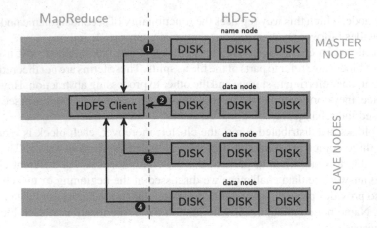

Fig. 6.4 HDFS read flow

6.2.3 Write Flow

The write flow is presented in Fig. 6.5. In step 1, name node creates a new file by request from the HDFS client. Name node is responsible for verifying client's permissions and checks for potential conflicts with already existing files. You probably noticed it many times trying to output results of your MapReduce program to already existing folder. If verification is successful, the client can start writing the data.

In step 2, the client writes data and as they are being processed new block allocations are requested from name node, both for the first copy of the block and the remaining replicas, each copy on a different node. However, the client is only

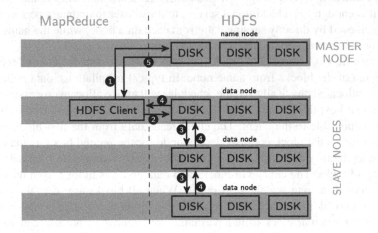

Fig. 6.5 HDFS write flow

responsible for the first replica. After that, in step 3, the data node on which the last replica is stored takes over and creates another replica based on the list of data nodes provided by name node. Each node acknowledges the operation in step 4.

When client finishes, writing it notifies the name node in step 5 and name node will wait for the minimum amount of replicas to be created before marking write as successful; this amount is by default set to 1. In case the writing of some of the replicas fails, name node will create additional replicas of underreplicated blocks to ensure the right amount is present.

6.2.4 HDFS Failovers

Failover data node. Data nodes periodically report to name node. If name node loses contact with a data node, it removes it from available resources and determines which data splits it was holding. Name node orders replication of the lost splits from the other existing copies. The whole process is transparent to the user.

Failover name node. Name node failure handling depends on the configuration of Hadoop. If traditional Hadoop 1.0 configuration is used with a secondary name node, then, in contrast with data node failure, name node failure is not transparent. A typical procedure is to use secondary name node and populate it with data from primary name node's metadata backup. Data on the secondary name node cannot be used directly because it is always a little behind the primary. This operation requires manual intervention.

If Hadoop 2.0 HDFS HA is used, then standby name node can step in immediately becoming active, and any issues with former active name node have to be fixed manually, but the operation of the cluster should remain undisturbed.

6.3 Job Handling

In this section, I look deeper in job handling for typical job flow. Then, I consider the issue of data locality and finally, I discuss procedures for handling typical job- and task-related failures. The focus of data between jobs in general rather than on details of MapReduce, as it is described in great detail in another chapter.

6.3.1 Job Flow

Figure 6.6 illustrates a complete job flow for a job with three map tasks and one reduce task. Data processing in the Hadoop framework is organized in jobs, which consist of three elements: input data, configuration, and program. The program is subdivided into tasks, which are composed in map and reduce phases.

Execution of job is controlled by JobTracker and TaskTracker. JobTracker decided which tasks run under which TaskTracker. TaskTracker is then responsible for monitoring the progress of each task assigned to it.

Hadoop divides input data into parts of equal size. These splits usually correspond one-to-one to blocks in HDFS, but it is possible to create splits consisting of several blocks. JobTracker provisions one map task for each split.

Increasing amount of splits increases parallelism, what means that large input can be processed by more map tasks at the same time. In principle, this should reduce processing time. In practice, there is a trade-off between the amount of map tasks, the size of each split, and computation requirements per split. Each map task spins off a new JVM creating significant overheads that reduce benefits of a very large amount of splits in the map phase.

JobTracker attempts to allocate map tasks to nodes that host respective blocks of data, which information can be obtained from HDFS' name node. This reduces network bandwidth use and as a result, improves processing time. Only when all nodes holding necessary blocks are already executing, maximum allowed amount of map tasks (which is configurable) would JobTracker schedule map task on a different nodes as close as possible to a data node containing necessary data.

In Fig. 6.6, you can notice that map tasks, in contrast, to reduce tasks, save data locally on each node, instead of HDFS. The output of map tasks is only required until the job completes and is then discarded. Storing it in HDFS would incur unnecessary performance penalty due to replication.

Reduce tasks are not allocated w.r.t. location of input blocks in HDFS. Input data for reduce tasks come from all output data from map tasks; as a result, data transfer between map and reduce stage happens over the network; this stage is called shuffle. As represented in the figure, reduce tasks typically store the results in HDFS. This result in network traffic equivalent to a regular HDFS write operation.

6.3.2 Data Locality

I mentioned data locality several times in this chapter. Let us address it directly now. Locality is respective location between map task and data split it has to process. The optimal situation is when they are collocated on the same physical node. If this is impossible, e.g., all task slots are already full on a particular node, in-rack locality is preferred over in data center locality.

Usually, the size of splits for map tasks corresponds to the size of a block in HDFS, which by default is 64 MB. The reason they are the same is that it is the most direct way for MR framework to guarantee data locality for the task. Block and split size can be large; 128 MB is also common, 512 MB not untypical. The optimal size depends on the task to perform. Changing block size will only affect new files written to the system. Changing split size affects individual jobs. With

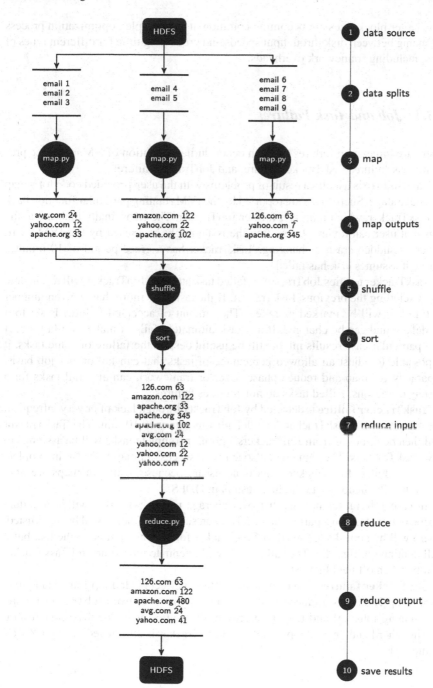

Fig. 6.6 Basic data flow during job execution

time larger block sizes are becoming common, it is a complex optimization process balancing between disk throughput speeds and processing time for different types of jobs, including framework overheads.

6.3.3 *Job and Task Failures*

There are three main failures that can occur during execution of a MapReduce program: task failure, TaskTracker failure, and JobTracker failure.

Task failure is usually a result of problems with the user-provided code for a map or reduce stage. Such errors are reported by the JVM running particular function back to TaskTracker, which reports it further JobTracker. The error finally makes it to the logs that user can review. Less often the issue might be caused by the JVM itself, either its sudden crash or hanging. TaskTracker has a grace period of 10 minutes before it assumes task has failed.

TaskTracker notifies JobTracker of failed task attempt. JobTracker will reschedule task excluding the previous TaskTracker. If the task fails more than a given number of times, it will be marked as failed. The amount of accepted failures is set to 4 by default and can be changed. If any task ultimately fails, it makes job fail too. If in a particular case results might still be useful despite the failure of some tasks, it is possible to adjust an allowed percentage of tasks that can fail on per job basis, separately for map and reduce phase. User or framework can also kill tasks for a variety of reasons. Killed tasks do not count as failures.

TaskTracker failure is detected by JobTracker through lack of or very infrequent heartbeat sent by TaskTracker. The default time is set to 10 min. The TaskTracker will then be removed from JobTrackers's pool. Incomplete tasks will be assigned to new TaskTrackers. The same will also apply to successful map tasks for incomplete jobs from failed TaskTracker. This is necessary because results of maps are only stored locally, in contrast to reduce results in HDFS.

In case JobTracker notices that over-average number of tasks within a certain period of time fails on a particular TaskTracker, such TaskTracker will be blacklisted. There will be no tasks assigned to TaskTracker for as long as it is on the list, but it will continue to run idle. The failure rate will attenuate with time and TaskTracker will be taken off the blacklist.

JobTracker failure has a completely different effect in Hadoop 1 and Hadoop 2. In Hadoop 1, there is no mechanism to gracefully recover from JobTracker failure. All running jobs fail and have to be resubmitted after manual JobTracker restart. On the other hand, in Hadoop 2, failure of JobTracker is eliminated through YARN framework.

6.4 Exercises

Exercise 6.1   10% of disks fail over 3 years, this is ca. 3% a year, and 0.008% a day. Assume each data node has 10 disks and you have 50 racks with 30 data nodes each. How many disks will fail on a typical day and week?

Exercise 6.2   Based on the calculations from Exercise 6.1, what is the probability of data loss with 1, 2, and 3 replicas? Assuming failed disks are replaced within a day.

Exercise 6.3   Imagine you bought half of the disks in your datacenter at the same time. Does it affect possible data loss calculated in Exercise 6.2.

Exercise 6.4   When executing a job you should optimize its performance. Three main factors to observe are data transfer from drive, data transfer from network, and processor activity. How will you recognize that your job might require further optimization looking at just these three parameters?

Chapter 7
MapReduce Algorithms and Patterns

In this chapter, I will show you a few examples of the most common types of MapReduce patterns and algorithms. They will guide your thinking on how to encode typical operations in a MapReduce way. This should guide you in a way you think about your own coding challenges.

All code in this chapter is written basing on MRJob.[1] This is a library that simplifies writing MR code for Hadoop in Python. It wraps around standard Hadoop Streaming but conceals most of its unnecessary complexity. MRJob provides you with a simple interface comparable with native Java programming for Hadoop.

You can find more examples of algorithms and patterns in dedicated books. Two good examples are "MapReduce Design Patterns" Miner and Shook (2012) and "Data-Intensive Text Processing with MapReduce" Lin and Dyer (2010). Please keep in mind that most examples you will find in these books or elsewhere on the Internet are coded in Java. However, with MRJob, it should be fairly straightforward to produce a Python equivalent.

The next five sections correspond to different types of patterns and algorithms, including (1) counting, summing, and averaging; (2) constructing a search assist index; (3) random sampling input dataset; (4) processing multiline inputs; and (5) creating an inverted index. They are all based on Apache Software Foundation Public Mail Archives, which you are already familiar with from several earlier chapters, in particular, Chap. 3. Through these examples, you will become familiar with techniques such as multistep jobs, map-only jobs, mapper initialization, and parameter passing.

---

[1] https://github.com/Yelp/mrjob.

© The Author(s), under exclusive license to Springer Nature Switzerland AG 2019
T. Wiktorski, *Data-intensive Systems*, Advanced Information and Knowledge Processing,
https://doi.org/10.1007/978-3-030-04603-3_7

7.1   Counting, Summing, and Averaging

Counting, in various forms, is one of the most common operations you will encounter in MapReduce. Most of the algorithms or more complex operations contain some form of counting pattern. Getting a good grasp of this basic operation is the groundwork that will pay off when you work on advanced problems.

I will start by showing you one of the most elementary examples adapted to the dataset you are already familiar with. You will count the number of emails sent from each domain. This is a very similar task to *word count* example, which you usually see in most Hadoop books. This email counting example is essentially the same as presented in Chap. 3. However, this time, I will go more into detail explaining its inner workings. In the second example, I will show you how to extract more complex statistics, such as maximum, minimum, and average. It will serve as a canvas to demonstrate a multistep MapReduce algorithm.

Listing 7.1   Counting amount of emails sent from each domain

```
1  from mrjob.job import MRJob
2
3  class MRCountSum(MRJob):
4
5      def mapper(self, _, line):
6          line = line.strip()
7          if line.find("From:") == 0:
8              email_domain = line[line.find("@")+1:line.find(">")]
9              if len(email_domain) == 0:
10                 email_domain == "empty"
11             yield email_domain, 1
12
13     def combiner(self, key, values):
14         yield key, sum(values)
15
16     def reducer(self, key, values):
17         yield key, sum(values)
18
19
20 if __name__ == '__main__':
21     MRCountSum.run()
```

The first example in Listing 7.1 uses the most typical one-step map–combine–reduce pattern. If, in addition, you also use standard naming convention mapper–combiner–reducer, you do not have to explicitly define the step. You will learn how to define the step in the next example. MapReduce program needs at least one mapper or one reducer to execute.

Map phase is defined between lines 5 and 11. Input is processed line by line and only lines starting with "From: " are taken in consideration, as filtered in line 7. The domain is then extracted in line 8 and finally, output key–value pair is defined in line 11. Key is the domain name and value is always 1.

Returning 1 as a value for some key is the standard MapReduce way to count things. Depending on your algorithm, it might also be a value other than 1. These values are then grouped by key in reducer in lines 16–17. In this case, we simply sum them all together, but other operations can also be applied. In this case, combiner, in lines 13–14, is the same as the reducer. Its goal is to sum as many 1's as early as possible, to minimize traffic from map to reduce phase.

The last thing I wanted to touch upon is the `yield` keyword, which you can see in each function we defined in this program. In simple terms, `yield` is like `return`, but it returns a *generator*. Generators, in this case, improve performance, but they also have further Pythonic implications. What you should remember is that this is the way to generate key–value pairs in MapReduce programs in Python.

Listing 7.2 Running counting job in three different modes

```
1 python count_sum.py -r inline hadoop.txt
2 python count_sum.py -r local hadoop.txt
3 python count_sum.py -r hadoop hdfs:///dis/hadoop.txt
4 python count_sum.py -r hadoop hdfs:///dis/hadoop.txt --output-dir
     hdfs:///dis/output1
5
6 hadoop fs -text /dis/output1/part* | less
```

Listing 7.3 Example output of count job

```
1 "aol.co"       1
2 "aol.com"      18
3 "aol.in"       3
4 "apache.or"    12473
5 "apache.org"   144144
6 "apache.org<mailto:cutting@apache.org" 1
7 "apache.org]=2" 49
8 "apics.co.u"   1
9 "apple.com"    47
```

In Listing 7.2, you execute the simple counting program in four different ways. First, you use the *inline* runner in line 1. This is the simplest runner, similar in its operation to the *standalone* mode in Hadoop. In line 2, you use the *local* runner, which is similar to the *pseudo-distributed* mode in Hadoop. In both cases, you start by executing `python`, followed by the name of the file with the MapReduce program, option `-r` specifies the runner. The last argument is the file name. In both cases, it is a file in a local filesystem.

Finally, in line 3, the job is executed on Hadoop. Two things that change are the runner option and the location of the file. You can see that we now specify a file in HDFS. By default, MRJob outputs results to *stdout*. You might also want to save the results for later use. To do so, you use option `-output-dir`, this can be both local and HDFS directory. We make use of this possibility in line 4. This will result in the result being both saved to the directory and sent to *stdout*. If you want just to save the results to a file, it is enough to add `-no-output True` to your command.

In order to read the results saved to an HDFS file, we can use `hadoop fs -text` command and pipe the output to `less` to make reading easier, especially for large outputs. This is demonstrated in line 6. A fragment of possible output is presented in Listing 7.3.

Tracking errors in MapReduce programs is difficult. You have to dig through many layers on Hadoop system and any additional libraries you might be using. For instance, if you forget to set the correct permissions, then the error messages you will receive will be of little help to narrow down the problem. This is why it is extremely important to always go through both *inline* and *local* testing before you execute your program on a Hadoop cluster.

Listing 7.4   Finding average, maximum, and minimum amount of emails per domain

```
 1  from mrjob.job import MRJob
 2  from mrjob.step import MRStep
 3
 4  class MRCountSumAvg(MRJob):
 5      def steps(self):
 6          return [
 7              MRStep(mapper=self.mapper_count,
 8                     combiner=self.combiner_count,
 9                     reducer=self.reducer_count),
10              MRStep(mapper=self.mapper_avg,
11                     reducer=self.reducer_avg)
12              ]
13
14      def mapper_count(self, _, line):
15          line = line.strip()
16          if line.find('From:') == 0:
17              email_domain = line[line.find('@')+1:line.find('>')]
18              if len(email_domain) == 0:
19                  email_domain == 'empty'
20              yield email_domain, 1
21
22      def combiner_count(self, key, values):
23          yield key, sum(values)
24
25      def reducer_count(self, key, values):
26          yield key, sum(values)
27
28      def mapper_avg(self, key, value):
29          yield 'amount', value
30
31      def reducer_avg(self, key, values):
32          values_list = list(values)
33          yield 'avg emails per domain',           sum(values_list)/len(values_list)
34          yield 'max emails per domain', max(values_list)
35          yield 'min emails per domain', min(values_list)
36
37  if __name__ == '__main__':
38      MRCountSumAvg.run()
```

I follow now with a more advanced example. We will calculate average, maximum, and minimum amount of emails per domain. First three steps, now called `mapper_count`, `combiner_count`, and `reducer_count`, are the same as in the previous example. However, instead of simply displaying or saving data from reducer, they are now directed to an extra step consisting of a new mapper and reducer.

Because now we have more than one step, both steps together with their elements have to be explicitly defined. This is done in lines 5–12. The first step with mapper, combiner, and reducer is defined in lines 7–9, and the second step with just mapper and reducer in lines 10–11.

I skip over the description of the code in lines 14–26, as you are already familiar with it. Let us focus on two functions in the second step. In the `mapper_avg` in line 29, we output all values with the same key because we want to calculate statistics across all domains. The value of the key does not matter. The amount of values we will output is not that large because they have already been aggregated by domain in the previous step.

In the reducer, we first take `values` variable and convert it to a list in line 32. This is a key operation. Remember that *yield* in map phase returned a generator. It is a property of generators that they can only be iterated over once, but we need to calculate four different values based on this generator. After the conversion to list, in line 33 we calculate and return average, in line 34 maximum, and finally in line 35 minimum. The order is arbitrary.

Listing 7.5 Running stats job in four different modes and displaying results

```
1 python count_sum_avg.py -r inline hadoop.txt
2 python count_sum_avg.py -r local hadoop.txt
3 python count_sum_avg.py -r hadoop hdfs:///dis/hadoop.txt
4 python count_sum_avg.py -r hadoop hdfs:///dis/hadoop.txt --output-dir
     hdfs:///dis/output2
5
6 hadoop fs -text /dis/output2/part* | less
```

Listing 7.6 Example output of stats job

```
1 "avg emails per domain" 125
2 "max emails per domain" 144144
3 "min emails per domain" 1
```

We run the program in four different ways in Listing 7.5 with example results in Listing 7.6. Please remember that output directory we define in line 4 of the former listing cannot already exist. This is a standard Hadoop behavior. You must either remover it before running your program or change the output directory. Remember to run the program in all three runner modes to debug as much as you can in simpler modes.

7.2 Search Assist

Predicting the consecutive word is a common need in many applications concerned with user input, e.g., search assist which suggests the next word you a query. It is also an important step in several areas of science such as computational linguistics or DNA sequencing. An algorithm that implements such functionality is fairly simple and, what is good for us, is a natural fit for a MapReduce approach.

In this section, we will construct a simple MapReduce program that creates a list of words and for each word provides an ordered list of most frequent successors.

This list can serve as a search assist index. It will be based on the subjects of email in the Apache archive, but it could also very easily be extended to the email body. Structure of the code will roughly be based on the two-step pattern from the previous section.

Listing 7.7 Search assist, top 3 following words

```
1  from mrjob.job import MRJob
2  from mrjob.step import MRStep
3
4  class MRSearchAssist(MRJob):
5      def steps(self):
6          return [
7              MRStep(mapper=self.mapper_count,
8                     combiner=self.combiner_count,
9                     reducer=self.reducer_count),
10             MRStep(mapper=self.mapper_trans,
11                    reducer=self.reducer_trans)
12             ]
13
14     def mapper_count(self, _, line):
15         line = line.strip()
16         if line.find('Subject:') == 0:
17             line1 = line[9:].split()
18             pairs = zip(line1, line1[1:])
19             for pair in pairs:
20                 yield pair, 1
21
22     def combiner_count(self, pair, count):
23         yield pair, sum(count)
24
25     def reducer_count(self, pair, count):
26         yield pair, sum(count)
27
28     def mapper_trans(self, pair, count):
29         first = pair[0]
30         second = pair[1]
31         yield first, (second, count)
32
33     def reducer_trans(self, word, second_list):
34         sorted_list = sorted(list(second_list), key = lambda second:
                            -second[1])
35         if len(sorted_list)>=3:
36             yield word, [second for (second, count) in sorted_list[:3]]
37
38 if __name__ == '__main__':
39     MRSearchAssist.run()
```

In the first step, lines 14–20, of Search Assist program in Listing 7.7, we follow the same steps as in the counting example, but in line 16 we look for the subject line instead of email address. The subject is tokenized into a list of single words in line 17. Pairs of consecutive words are created in line 18 by *zipping* the list with itself. The pairs are output in lines 19–20 using standard counting pattern.

Counts for pairs are summed first in the combiner in lines 22–23 and finally in the reducer in lines 25–26.

In the map phase of the second step, lines 28–31, we split each pair and output it in a new arrangement. The first word in the pair becomes a key, you can think of it as the word the user is typing in. Value becomes a tuple where the first element is the consecutive word, and the second element is a number of an amount of occurrences of the second word after the first one.

All the values arrive at the reducer, in lines 33–36, grouped by the first word and for each, we now have a list of consecutive words with amounts of occurrences. We sort this list in a descending order by the number of occurrences in line 34. Finally, for words that have at least three successors, line 35, we output three most common successors in line 36. The number three was chosen for the sake of readability of the result and you could simply pick a different one.

You probably noticed that we did not use the combiner in the second step. Including it would be code less intuitive, because it would require changes both to mapper and reducer. As you remember, the output of combiner has to be the same as the output of mapper. With a simple mapper as ours here, this is not possible. The requirement on the output of mapper is important, because it is up to MapReduce framework to decide if to run combiners at all. Your code has to be able to handle both scenarios, with and without combiners executed.

Listing 7.8 Running search assist in four different modes and displaying results

```
1 python search_assist.py -r inline hadoop.txt
2 python search_assist.py -r local hadoop.txt
3 python search_assist.py -r hadoop hdfs:///dis/hadoop.txt
4 python search_assist.py -r hadoop hdfs:///dis/hadoop.txt --output-dir
      hdfs:///dis/output3
5
6 hadoop fs  cont /dis/output3/part* | less
```

Listing 7.9 Example output of search assist job

```
1 "wrongly"  ["sets", "log", "classifies"]
2 "wrt"   ["DFS", "security", "missing"]
3 "xml"   ["files", "string", "file"]
4 "year"  ["for", "to", "field"]
5 "yet"   ["guarantees", "task", "to"]
6 "you"   ["run", "an", "rather"]
7 "your"  ["project", "job", "own"]
8 "zero"  ["size", "reduces-", "map"]
9 "zip"   ["files", "file", "file?"]
```

We run the program in four different ways in Listing 7.8 with example results in Listing 7.9. Please remember that output directory we define in line 4 of the former listing cannot already exist. This is a standard Hadoop behavior. You must either remove it before running your program or change the output directory. Remember to run the program in all three runner modes to debug as much as you can in simpler modes.

7.3   Random Sampling

We use MapReduce when we need to process large datasets. However, sometimes we might want to test our programs just on a sample selected from this large dataset. In this section, we construct a simple program that extracts a random sample of subjects from all emails.

It gives us an opportunity to learn three new patterns. This will be our first map-only job that you heard about earlier in the context of *embarrassingly parallel* problems. We will also use a new type of function `mapper_init`, which is an addition to the standard mapper, combiner, and reducer set you already know. Finally, you will learn how to pass an argument to the job from the command line.

Listing 7.10   Random sampling email subjects with given probability and seed

```
1  from mrjob.job import MRJob
2  import random
3
4  class MRRandomSampling(MRJob):
5
6      def configure_options(self):
7          super(MRRandomSampling, self).configure_options()
8          self.add_passthrough_option('--samplingRate', type='float', default=.1)
9          self.add_passthrough_option('--seed', type='float', default=10)
10
11     def mapper_init(self):
12         random.seed(self.options.seed)
13
14     def mapper(self, _, line):
15         line = line.strip()
16         if line.find('Subject:') == 0:
17             if random.random()<self.options.samplingRate:
18                 yield _, line[9:]
19
20 if __name__ == '__main__':
21     MRRandomSampling.run()
```

In the Listing 7.10, we start by introducing a new element `configure_options` in lines 6–9. In it, we define two command line options we want to pass to our program. First, option in line 8 is `samplingRate` and it determines a relative size of the sample w.r.t. the whole dataset. In this case, we do not explicitly mean sampling rate as understood in signal processing, but we use it rather as a layman's term for probability in random sampling. The difference is only theoretical if we consider just the final output size, but it has a practical impact on which samples are in fact selected.

The second option in line 9 is `seed` and it allows us to define the seed for the random number generator. This way we can get repeatable results when necessary.

Random number generator is seeded in line 12 as a part of the `mapper_init` function in lines 11–12. This function is run once on each node before the map phase begins, in contrast to the `mapper` function which is run for each input in the map phase. It is crucial that the seed is set in the initialization. If it was set in the `mapper` than it would be reset, in this case, at every topic line and we would get the same *random* number each time. This is obviously incorrect. Setting the same speed and sampling rate will provide you with the same result on the same cluster, but not necessarily on different clusters. There result will also be different in `inline` and `local` running modes.

Finally, in mapper in lines 14–18, we extract the subject in line 16 and proceed to output it w.r.t. given probability in line 17. When constructing the actual output in line 18, we only emit the value, omitting the key.

Listing 7.11 Running random sampling in four different modes and displaying results

```
1  python random_sampling.py --seed 5 --samplingRate .0001 -r inline hadoop.txt
2  python random_sampling.py --seed 5 --samplingRate .0001 -r local hadoop.txt
3  python random_sampling.py --seed 5 --samplingRate .0001 -r hadoop
       hdfs:///dis/hadoop.txt
4  python random_sampling.py --seed 5 --samplingRate .0001 -r hadoop
       hdfs:///dis/hadoop.txt --output-dir hdfs:///dis/output4
5
6  hadoop fs -text /dis/output4/part* | less
```

Listing 7.12 Example output of random sampling job

```
1  null    "[jira] Commented: (HADOOP-6194) Add service base class and tests to"
2  null    "svn commit: r508622 - in /lucene/hadoop/trunk: CHANGES.txt"
3  null    "svn commit: r477423 - in /lucene/hadoop/trunk: CHANGES.txt"
4  null    "Problem building a inputformat using 0.18.1"
5  null    "RE: my company is hiring engineers"
6  null    "[jira] Updated: (MAPREDUCE-1033) Resolve location of configuration"
7  null    "[jira] Created: (HDFS-1694) SimulatedFSDataset changes to work with"
8  null    "[jira] Commented: (HDFS-941) Datanode xceiver protocol should allow"
9  null    "svn commit: r791889 - /hadoop/hdfs/site/"
10 null    "[jira] Updated: (HADOOP-1601) GenericWritable should use"
11 null    "[jira] Commented: (HADOOP-3863) Use a thread-local rather than"
12 null    "[jira] Commented: (HADOOP-1905) Abstract node to switch mapping"
13 null    "[jira] Updated: (HADOOP-5441) HOD refactoring to ease integration"
14 null    "[jira] Commented: (HADOOP-2068) [hbase] RESTful interface"
```

We run the program in four different ways in Listing 7.11 with example results in Listing 7.12. We pass seed as `-seed` and sampling rate as `-samplingRate`; all other options remain the same as in previous examples.

Please remember that output directory we define in line 4 of the former listing cannot already exist. This is a standard Hadoop behavior. You must either remove it before running your program or change the output directory. Remember to run the program in all three runner modes to debug as much as you can in simpler modes. I encourage you to experiment with different seeds and sampling rates.

7.4 Multiline Input

In many cases, you might need to process an input in a block that spreads across many lines of text. One way to approach it would be to write your own input format. However, such an approach requires special knowledge and additional effort. As a result, it might not be worth pursuing it. It might be easier to continue processing the input line by line and find a workaround to adapt it to a multiline input. We will try such an approach here. For simplicity, we omit issues related to how the input is divided into splits, which might come into consideration in a production setting.

In this section, we will again focus on a map-only job, which, as you remember, means it is embarrassingly parallel. This is what MapReduce is best at. We will use `mapper_int` to maintain variables between `mapper` calls done for each line of input. We will also practice having an advanced `mapper` section.

Listing 7.13 Processing inputs with multiple lines

```
1  from mrjob.job import MRJob
2
3  class MRMultilineInput(MRJob):
4      def mapper_init(self):
5          self.message_id = ''
6          self.in_body = False
7          self.body = []
8
9      def mapper(self, _, line):
10          line = line.strip()
11          if line.find('Message-ID:') == 0:
12              self.message_id = line[13:len(line)-1]
13
14          if not line and not self.in_body:
15              self.in_body = True
16
17          if line.find('From general') == 0 and self.in_body:
18              yield self.message_id, ''.join(self.body)
19              self.message_id = ''
20              self.body = []
21              self.in_body = False
22
23          if self.in_body:
24              self.body.append(line)
25
26  if __name__ == '__main__':
27  MRMultilineInput.run()
```

In the Listing 7.13, we start by defining three general variables in `mapper_init` in lines 4–7. We will use `message_id` to hold ID of a current email message we are processing; `in_body` is a helper variable to help us work through multiple lines. Finally, `body` will hold contents of email body. Notice that it is defined as a list and not a string; we will discuss the reason for it later.

The `mapper` has four main parts. In the first part, lines 10–12, we clean the input line and check if it contains message ID. If so, we store the ID in the respective variable.

In the second part, lines 13–14, we check if the line is empty. If it is and we are not currently processing the body of the email, then we set helper variable `in_body` to `True`. This method is based on an observation that body of the email in the ASF Public Mail Archives is separated from the head of the email by an empty line.

In the third part, lines 16–20, we look for the phrase "`For general`'?, which is a sign of a beginning of a new email. When we encounter it, we perform a set of operations. First, we output a key–value pair with message ID as the key and body of the email as a value. Then, we reset all three helper variables.

In the last fourth part, lines 22–23, we simply add a line to the body variable in case we are in the body. Notice that in fact all these four parts are independent and only one is effective for each line of the email we process.

I mentioned earlier that this map-only job is a great example to explore scalability of MapReduce. Nevertheless, we should be careful and optimize the inner workings of our algorithms not to waste the effects of scalability. The performance in this example could be easily destroyed by inefficient string processing of the email body; this is why we use an array. Such issues are particular to language, its version, and often even a version of the framework. In the case when the performance is much worse than expected, especially in the map phase, you should inspect your code for similar issues.

Listing 7.14 Running multiline input processing

```
1 python multiline_input.py -r inline hadoop.txt
2 python multiline_input.py -r local hadoop.txt
3 python multiline_input.py -r hadoop hdfs:///dis/hadoop.txt
4 python multiline_input.py -r hadoop hdfs:///dis/hadoop.txt --output-dir
    hdfs:///dis/output5
5
6 hadoop fs -text /dis/output5/part* | less
```

Listing 7.15 Example output of multiline input job

```
1 "4BB22BFE.30906@apache.org" "Andy Schlaikjer wrote:> Can anyone comment on why
    the base *Writable implementations (e.g.> IntWritable, LongWritable, etc)
    don't specify WritableComparable<T>'s> type param T?I suspect it's just
    historic. When they were first written we were notusing Java generics, and
    no one has ever retrofitted them with generics.Doug"
2 "7c0da9a1003300701o5e7f8a2ifd4a2d262bf92651@mail.gmail.com" "I've received a
    number of responses from helpful Hadoop list members. Thanks!But I should
    clarify here; I'm not looking for workarounds for my Pairdeclaration or
    explanations of Java's generics facilities. I'm lookingfor justifications
    for Hadoop's approach.Looking at Java's core library, there are 86 classes
    which implementjava.lang.Comparable<T> [1]. 80 of these specify T = the
    type beingdeclared, or some parent of the type being declared.[1]
    http://java.sun.com/javase/6/docs/api/java/lang/Comparable.htmlSo, I'm
    curious, are there specific use cases the Hadoop community hasrun into
    which support the current design of the primitive *Writabletypes which
    excludes specification of Comparable<T>'s T param? If notI'll try and find
    time to work on a patch to push more conventionaluse of generics into
    Hadoop Common.Best,"
```

We run the program in four different ways in Listing 7.14 with example results in Listing 7.15. Please remember that output directory we define in line 4 of the former listing cannot already exist. This is a standard Hadoop behavior. You must either remove it before running your program or change the output directory. Remember to run the program in all three runner modes to debug as much as you can in simpler modes.

7.5 Inverted Index

The inverted index provides a mapping from content to location. It is similar to a concordance or index you find in the back of many books. It tells you for each important word or phrase on which pages it occurs. For centuries, it was constructed manually and could take years to complete. The inverted index is one of the components of indexing algorithms used by search engines.

For the first time, instead of our standard input file, we use the output of an earlier algorithm as an input. To make it possible, we introduce protocols as a way to serialize output and deserialize input in different ways.

Listing 7.16 Calculating inverted index

```
1  from mrjob.job import MRJob
2  from mrjob.protocol import JSONProtocol
3
4  class MRInvertedIndex(MRJob):
5      INPUT_PROTOCOL = JSONProtocol
6
7      def mapper(self, email_id, email_body):
8          for word in email_body.split():
9              if email_id:
10                 yield word, email_id
11
12     def combiner(self, word, email_ids):
13         email_ids_set = set(email_ids)
14         for email_id in email_ids_set:
15             yield word, email_id
16
17     def reducer(self, word, email_ids):
18         email_ids_list = list(set(email_ids))
19         yield word, (len(email_ids_list),email_ids_list)
20
21 if __name__ == '__main__':
22     MRInvertedIndex.run()
```

In Listing 7.16, we start by defining INPUT_PROTOCOL in line 5. The default input protocol is just raw text, which is often what you want as we have seen in all earlier exercises. MRJob uses JSONProtocol as default for all protocol for output and, in fact, all data transfers except the input. We change the input protocol in this case because we use the output of an earlier job as the input.

In mapper, lines 7–10, we first split the body of the email into separate words and then output each word as key and email ID as body. This order comes from the fact that index is to be inverted.

In combiner, lines 12–15, data receive data grouped by word. For each word, we convert the input list email_ids to a set; this way, we eliminate any duplicates. We then output each word and email ID in the same way as in the mapper. By eliminating duplicate, we can save bandwidth between map and reduce phase. Notice that we do not keep counts of each email ID for each word, but we could if we later wanted to introduce some measure of importance based on frequency.

In `reducer`, lines 17–19, we again use a set to eliminate duplicates. At this stage, the duplicates might result from merging data from different `mappers` or from `combiners` not being executed at all. This is always a possibility we have to account for. We also change the set to the list right away, this is to match the types of data a protocol can represent. Not all value types can always be represented by every protocol. A set is one such example and our choice is either to change the protocol or change the type. Finally, we output word as the key and a tuple consisting of list length and the list itself as the value.

Again, I would like to draw your attention to the processing of generators in `combiner` and `reducer`. We already discussed that generators can only be iterated over once. This can lead to unexpected results and errors if combined with `yield` statement, such issues might be very difficult to debug.

Listing 7.17 Running inverted index

```
1 python inverted_index.py -r inline email_id_body_1m.txt
2 python inverted_index.py -r local email_id_body_1m.txt
3 python inverted_index.py -r hadoop hdfs:///dis/output6
4 python inverted_index.py -r hadoop hdfs:///dis/output6 --output-dir
    hdfs:///dis/output7
```

Listing 7.18 Example output of inverted index

```
1 "APIs" [72,
       ["219D8244D980254ABF28AB469AD4E98F03469FCB@VF-MBX13.internal.vodafone.com",
       "AANLkTina2dxXd61b5zj4ZpS6-MK3jpzx47ecc3xZkP-r@mail.gmail.com", ...
2 "wind" [1, ["4ACA618F.5040709@lifeless.net"]]
3 "wrote:>" [1713, ["4BD95145.8040204@apache.org",
       "49A4E262.7080508@taragana.com",
       "AANLkTinPH9WeK_0mSo7FVUFQdFos0M60H80KaF6VU=Ny@mail.gmail.com", ...
4 "your" [981, ["1eabbac30912311118n484c1e73xf6139dace6f3ee27@mail.gmail.com",
       "ABC24175AFD3BE4DA15F4CD375ED413D0607D373B1@hq-ex-mb02.ad.navteq.com",
       "C7B199C8.358D%awittenauer@linkedin.com",
       "2a31deca1001131109s3386916ne4c8f6a45b02ad96@mail.gmail.com", ...
```

We run the program in four different ways in Listing 7.17 with example results in Listing 7.18. Please remember that output directory we define in line 4 of the former listing cannot already exist. This is a standard Hadoop behavior. You must either remove it before running your program or change the output directory. Remember to run the program in all three runner modes to debug as much as you can in simpler modes.

7.6 Exercises

Exercise 7.1 You probably noticed that output from the simple email counting program in Sect. 7.1 is rather unclean. Domains are sometimes incorrectly recognized or misspelled. Improve the program from Listing 7.1 to eliminate the incorrect domain names.

Exercise 7.2   Extend the search assist algorithm from Sect. 7.2 to calculate n-grams of any given length.

Exercise 7.3   Modify the search assist algorithm from Sect. 7.2 to include combiner in the second step.

Exercise 7.4   In the `reducer_trans` in Listing 7.7 if you changed the output to `yield word, second_list` you would get an error. Explain why?

Exercise 7.5   In random sampling Listing 7.10 move the procedure of setting seed from `mapper_init` to `mapper`. Observe how it changes the results.

Exercise 7.6   Rewrite random sampling example from Sect. 7.3 to get an evenly distributed fixed amount of samples from the dataset, instead of sampling with probability.

Exercise 7.7   Combining what you learned in Sects. 7.2 and 7.4 write a program that creates a search assist index based on the body of emails.

Exercise 7.8   In the multiline example from Sect. 7.4 change, the way body is processed, using string and += instead of a list. Observe the result.

Exercise 7.9   Extend the inverted index example from Sect. 7.5 to only return words that appear in more than 10 emails. Further, extend this example to only return three emails IDs for which this word appears most frequently. Include this frequency in the output and also use it to sort the email IDs.

Exercise 7.10   Modify the inverted index example from Sect. 7.5 to output a set instead of a list.

References

Lin J, Dyer C (2010) Data-intensive text processing with mapreduce. Morgan and Claypool Publishers. ISBN 1608453421, 9781608453429

Miner D, Shook A (2012) MapReduce design patterns: building effective algorithms and analytics for hadoop and other systems. O'Reilly Media, Inc

Chapter 8
NOSQL Databases

In chapters so far, you have relied on HDFS as your storage medium. It has two major advantages for the type of processing we desired to do. It excels at storing large files and enabling distributed processing of these files with help of MapReduce. HDFS is most efficient for tasks that require a pass through all data in a file (or a set of files). In case you only need to access a certain element in a dataset (operation sometimes called point query) or a continuous range of elements (sometimes called range query), HDFS does not provide you an efficient toolkit for the task. You are forced to simply scan over all elements to pick out the ones you are interested in.

Many so-called NOSQL databases offer solutions to this problem by employing a combination of different data models and indexing techniques. In the first section, we look into the development of NOSQL concept, including CAP and PACELC theorem. I also provide you with examples of the main types of NOSQL databases. Further, we explore HBase in greater detail, as one of the common NOSQL databases.

8.1 NOSQL Overview and Examples

For around a decade, we have observed a shift in the development of database technologies. Traditional relational database management systems based on ACID principles (Atomicity, Consistency, Isolation, Durability) have been giving place to alternative solutions. Since development and popularity of relational databases have been for long time tied with success of Structured Query Language (SQL), the newly developed databases were initially called NoSQL. Nevertheless, the key difference was not the lack of support for this language, but rather relaxation of ACID principles. The main goal was to increase performance and ability to handle large datasets in a distributed manner.

With time, it became apparent that SQL as a language, or its subsets, might be useful for any database; also a NoSQL database, which lead to capitalization of O

© The Author(s), under exclusive license to Springer Nature Switzerland AG 2019   77
T. Wiktorski, *Data-intensive Systems*, Advanced Information and Knowledge Processing,
https://doi.org/10.1007/978-3-030-04603-3_8

to shift the meaning from *no* to *not only*. Nevertheless, the key differences are a departure from ACID principles and alternative approach to the trade-off between consistency and availability, as explained in Sect. 8.1.1.

In NOSQL context, a word *store* is often used to describe a simple database with basic functionality.

Typical types of NOSQL databases include key–value, wide-column, document, object, and time series. To a certain extent, they are all a form of key–value databases with a varying degree of built-in support for querying and indexing over the value part of the key–value duo.

In a typical key–value database (e.g., Voldemort), the key is the only concern for the database. The user can only query based on the value of the key and will get the related value returned. The value part is not interpreted by the database.

The wide-column databases (e.g., HBase, Cassandra) also rely on the key as the only way to find records but they divide the value into columns. It allows to retrieve only parts of the record and to improve efficiency for scenarios where ranges of value for only a single column are necessary. It is a typical case for time series databases (e.g., OpenTSDB) that relay often on a wide-column database for actual storage.

Document databases (e.g., MongoDB) allow for a limited indexing and querying over elements of the value. It can be very useful for retrieving JSON files, but also for other applications.

Finally, object databases (e.g., Amazon S3) are useful for storing large, complex, varying data elements. In Cloud Computing context, they often replace file systems for long-term storage.

8.1.1   CAP and PACELC Theorem

When thinking about various NOSQL databases, regardless of the particular data model, it is useful to consider how they approach issues of availability and consistency of data. CAP theorem described by Gilbert and Lynch (2002), also referred to as Brewer's theorem, describes the relation between consistency, availability, and partition tolerance. The theorem states that for distributed databases it is possible to only guarantee two of the three of the aforementioned properties at the same time. For example, if a database guarantees consistency and availability, it cannot be partition tolerant.

Since partition tolerance has to be assumed for most NOSQL databases, the real choice is between consistency and availability. It limits the usefulness of the theorem. As a consequence, another theorem called PACELC was formulated by Abadi (2010). It states that in case a partition (P) is present, system has to choose between availability (A) and consistency (C). Else (E), when these are no partitions, system has to choose between latency (L) and consistency (C).

Thinking about database choice in the context of these theorems is crucial to ensure the expected behavior of the final application you are developing.

8.2 HBase Overview

HBase is not a replacement for an RDBMS in the Hadoop ecosystem. It is different in two ways from what people typically consider a database. First, it is not relational and it does not conform to ACID properties, which are central to relational databases but make distributed processing difficult. Due to the relaxing of these constraints, HBase can handle data in a distributed way efficiently.

The second difference is the fact that HBase is column-oriented, while most relational databases are row-oriented. Column and row orientation describe the way data are stored on the storage medium. In a row-oriented database, data from consecutive fields in the same record are stored one after another. In contrast, in a column-oriented database, data from the same field from consecutive rows are stored one after another.

If this is the first time you meet this concept, it might be easier to understand it with an example. Let us take our email dataset and focus on three fields: send date, sender's address, and topic. In a row-oriented database, value of send date for the first email would be followed by sender's address for that email and then the topic. After the whole first email, send date for the second email would follow, etc. On the other hand, in column-oriented database, values of send date for all emails would be stored first, then the sender's addresses for all emails, etc.

These are two drastically different data arrangements and they have different applications. Column-oriented databases are particularly efficient at range queries on column values because values from columns are collocated and can be read without unnecessary overhead. Efficient range queries are a basis for efficient data analysis. A good example of HBase application is various forms of time series data, which naturally fit the column-oriented approach. Calculating aggregates for different time periods, which is the most common use of time series data, naturally translates to range queries.

One might fairly ask whether these tasks could not be solved simply with a special columnar file format; two examples commonly used in Hadoop would be RCFile[1] and Parquet.[2] The answer is that these formats are optimized for different applications. Parquet focuses only on range queries and cannot support point queries. It is in principle read-only, meaning that in case of user would like to modify data a whole new file has to be created. It is obviously not optimal for continuous data storage. RCFile, despite some differences, shares the same general limitations. Nevertheless, they are both often used in Hadoop. RCFile is the common way to store data used in Apache Hive and Parquet in Cloudera Impala. In many applications, these file formats are interchangeable with each other, but cannot provide the functionality of HBase.

---

[1] https://cwiki.apache.org/confluence/display/Hive/RCFile.
[2] https://parquet.apache.org.

8.3   Data Model

Data in HBase is organized in tables, which are made of rows and columns. Cells are versioned automatically by timestamp and you can query data from a particular time. This feature is, however, not often used. Cell contents are byte arrays and are not interpreted by HBase. Row keys are also byte arrays and are byte-ordered. Data is accessed only by this primary key. Key construction is, therefore, very important. Values in columns are returned when you ask for a key, but you cannot query or filter by those columns or cells.

Columns are grouped into column families, distinguished by a common prefix. Column families have to be specified up front, but members of families can be added later, e.g., temperature:inside, temperature:outside. Columns in the same column family are stored jointly in the file system. It improves performance when fetching these values. When you decide on the particular data model for your application, you should organize column families with that in mind. As an effect, we could call HBase a column-family-oriented, and not column-oriented as it is often done.

Bloom filter can be enabled per column family to improve latency. Bloom filter, in great simplicity, is a lossy hash data structure with no false negatives but with true positives. Hash is populated with values from a set, but the size is constant. The more the elements, the higher probability of true positives. Bloom filter improves the performance of finding elements in the set by minimizing disk scans. The Bloom filter in HBase can be set per row or per row–column combination. Because columns, in fact, column families, are stored separately choosing between row and row-column, Bloom filter will offer different performances depending on your data structure and typical queries.

8.4   Architecture

In this section, I will introduce the architecture of HBase and provide more details on some of its elements. I will also discuss failover handling.

HBase consists of two main types of servers: HMaster and Region servers. HMaster is the master server that coordinates work of region servers, assigns regions to servers, and is responsible for basic operations such as creating and deleting tables. Region servers are responsible for storing and serving region data assigned to them.

In addition, ZooKeeper is used for coordination, handling tasks such as maintaining status through heartbeats. ZooKeeper is shared among all Hadoop ecosystem components.

8.4.1 Regions

The region contains a set of table rows. The number of regions grows as the table grows. Regions are the primary distribution units for an HBase cluster; in the same way, blocks are primary distribution units for an HDFS cluster. Data from regions reside in some underlying file system. In a typical scenario, this file system would be HDFS. Other files systems, e.g., local, can also be used.

Region server is responsible for interacting with the file systems and supporting read and write operations for any application using HBase. In a typical scenario, if region servers are collocated with data nodes, we know that one replica of data will be local to the region server.

8.4.2 HFile, HLog, and Memstore

To efficiently store data, HBase uses a special file format, called *HFile*. It is based on SSTable from Google's BigTable. Due to efficiency concerns, changes are written periodically to the file. It improves performance during write and read operations. The format is independent of the storage medium, but in practice, HDFS is used most often. Files in HDFS are immutable and it means that each write creates a new file. This is an important consideration for handling HFile writes.

Changes are accumulated in two different places: *WAL* (Write Ahead Log) and *memstore*. Memstore holds all modifications in memory and it is used to serve data for all the updates until they are flushed to HFile in the filesystem.

Memstore is efficient, but due to using a volatile medium as storage it is also risky. To mitigate this risk, data are first written to WAL to protect them against loss. WAL used by HBase is, unsurprisingly, called *HLog*. HLog is organized differently than HFile or memstore; it stores a chronologically organized list of changes. There is one HLog per region server instance. Data from HLog are only used in case of region server crash to replay lost changes.

8.4.3 Region Server Failover

As you might have already noticed, there is no region replication in HBase. Each region is handled by exactly on region server. This is a contrast to HDFS where each block usually has three copies, each at different data nodes.

To protect against data loss, e.g., in case of hardware failure, data has to be stored on a system that provides such protection. Usually, HBase is used in tandem with HDFS, but this is not the only option. You can even store data locally on your laptop and protect against data loss manually. In the standard HBase/HDFS tandem if region server goes down, the master will assign all regions from that region server to remaining region servers pointing them to the location of the file in HDFS.

Such recovery is possible because region server is only responsible for handling requests to data, but they do not actually store data. Typically, region server would be collocated with a data node, which holds a copy of region data locally.

What might be a bit confusing is that term *replication* is used in the HBase context. However, it means something different, then for HDFS. For HBase, it is a replication of the whole database, typically to a different location, a not of regions within the same database.

8.5   MapReduce and HBase

In this section, we go through a simple example, starting with loading data from HDFS to HBase and then running point and range queries on these data.

8.5.1   Loading Data

We start HBase by calling `hbase shell`. We create a new table called `hadooplist` with three column families. You can see the code in Listing 8.1.

Listing 8.1   Create HBase table

```
1 hbase shell
2 create 'hadooplist', 'sender', 'list', 'subject'
```

After the table is created, we can load the data from a CSV file. The easiest way to do it from the command line level is demonstrated in Listing 8.2.

Listing 8.2   Import CSV file to HBase

```
1 hbase org.apache.hadoop.hbase.mapreduce.ImportTsv
      -Dimporttsv.separator=,
      -Dimporttsv.columns="HBASE_ROW_KEY,list,sender,subject" hadooplist
      hdfs:///dis_materials/hadooplist.csv
```

8.5.2   Running Queries

If you are using standards *TableInputFormat*, one map task per region will be allocated. Usually, it gives good and even performance. Some performance issues can occur in case distribution of data between regions has a bias. In such a case, you might observe lower performance as some map tasks will slow down the rest. It can be dealt with using forced splits.

In Listing 8.3, you can see a simple point query, where we request data from an email with a particular ID. Thanks to HBase architecture, it is not necessary to scan the whole table to find one email and result is returned almost immediately. You can expect to maintain such performance even when the size of your tables grows.

Listing 8.3 Point query in HBase

```
1 get 'hadooplist',
      '20101021233704-FDE1E4F4-D765-45F8-B529-47166881AB95@linkedin.com'
```

In Listing 8.4, you can see three examples of range queries. In line 1, we run a simple range query without any conditions on the key but limiting the number of results to 5. In line 2, we ask for all email data from October 2, 2010 to October 5, 2010. In line 3, we define the same range as in line 2, but only ask for information about the sender of the email. You can notice that range queries are similarly fast as point query, as they take advantage of the same architectural advantages.

Listing 8.4 Range query in HBase

```
1 scan 'hadooplist', {LIMIT => 5}
2 scan 'hadooplist', {STARTROW => '20101002', ENDROW => '20101005'}
3 scan 'hadooplist', {STARTROW => '20101002', ENDROW => '20101005',
4      COLUMNS => ['sender:']}
```

However, running MapReduce jobs on data stored in HBase has also drawbacks. For data stored directly in HDFS, there are typically at least three available task trackers that can handle data locally. In case of HBase, MapReduce has access to only one available region server per region (the equivalent of split for task trackers). The choice between HDFS and HBase will depend on the data model of the use case you are working with, for instance, batch analysis vs random/range access.

HBase does not support SQL and there is no JOIN operation available either, as it is not a replacement for an RDBMS. HBase focus is on scanning over a large dataset based on the value of the key. If you would like to run SQL queries or join tables, you might take a look at Hive.

8.6   Exercises

Exercise 8.1   Extend import from Listing 8.2 to include more fields. Notice that you will need to create a new CSV file or use an alternative method to import data.

Exercise 8.2   Write a range query that includes both data and domain from which emails were sent.

Exercise 8.3   In the context of you project, think of at least two use cases one for HBase and one for another NOSQL database.

References

Abadi, D (2010) DBMS musings: problems with CAP, and Yahoo's little known NoSQL
    system. http://dbmsmusings.blogspot.com/2010/04/problems-with-cap-and-yahoos-little.html
    (visited on 09/26/2018)
Gilbert S, Lynch N (2002) Brewer's conjecture and the feasibility of consistent, available,
    partition-tolerant web services. en. In: ACM SIGACT News 33.2 (June 2002), p 51.
    ISSN: 01635700. https://doi.org/10.1145/564585.564601. http://portal.acm.org/citation.cfm?
    doid=564585.564601 (visited on 09/26/2018)

Chapter 9
Spark

9.1 Motivation

Functionalities of Hadoop (or other MapReduce frameworks) and Spark might seem duplicating, particularly from a high-level perspective. This is not surprising, considering that Spark is ultimately a type of MapReduce framework. However, Spark has gone a long way from a standard Hadoop stack of HDFS storage combined with the map and reduce routines. Development of Spark resulted, on the one hand, from important functionalities that were missing in Hadoop, on the other hand, from progress in hardware development that made some of the new functionalities feasible.

With every year since Hadoop conception, there was a growing need for an extension of the HDFS + MapReduce stack. The limitations are mostly related with exclusively disk-based storage and statelessness of data between MapReduce phases. All outputs are written directly to disk and all inputs read back from disk, replication further lowering the performance. Traditional MapReduce does not maintain files in memory between phases and jobs. Such an approach can be seen as a consequence of the literal implementation of functional data processing abstraction, but ultimately it a result of simplified scheduling in early Hadoop versions. It is important to notice that such simplification has little influence on ETL and basic analysis jobs, which Hadoop was originally used for. With the popularization of the paradigm and new applications of the framework—such as machine learning and advanced data analysis—this limitation became a major concern.

A prevailing hardware trend in the recent years is decreasing price of memory. Spark takes an advantage of this trend and maintains a subset of data in memory to improve performance. Disks, in some form, are still used to provide long-term non-volatile storage; in many applications, it is HDFS that takes this role. It is both due to legacy reasons—since Spark is deployed on top of existing Hadoop clusters—but it also improves the performances of the initial data loading process. The constantly decreasing price of memory together with also decreasing the price of fast interconnect leads to the separation of data and storage. This means a partial reversal of the

T. Wiktorski, *Data-intensive Systems*, Advanced Information and Knowledge Processing, https://doi.org/10.1007/978-3-030-04603-3_9

trend started by MapReduce and Hadoop, but only for long-term storage. Short-term storage directly related to data analysis remains localized.

The advanced data model in Spark supports a wider set of core operations. MapReduce provides in principle just the map and reduce function; available frameworks extend it usually just with simple performance extending routines (such as combiners). Other operations are not part of the core library and require additional libraries; Pig is probably the best example. Such an approach makes learning curve steep. Furthermore, with the advancement of Data Science, reliance on Java as basic programming language becomes inadequate. Hadoop Streaming is usually inconvenient to use for larger jobs and existing Python libraries have significant performance costs. There is also a need for a more interactive interface to support data exploration. Hive and HBase provide some form of such an interface, but they rely on additional data models that many application cannot conform to.

Finally, Spark is accessible from the two most important languages for Data Scientists—Python and R.

9.2   Data Model

The core of Spark's data model is Resilient Distributed Datasets (RDDs). Spark also provides DataFrames, which are a higher level structure built on top of RDDs; they require a stricter data model than RDDs. I will start by explaining the essence of RDDs and then move on to DataFrames. Finally, I will mention other available data structures and their applications. The practical use of these structures is exemplified in further sections of this chapter.

9.2.1   Resilient Distributed Datasets and DataFrames

RDDs are built as a set of manipulation originating from raw data. Data often come from HDFS, but can also come from other sources. Manipulations are organized as a DAG (Direct Acyclic Graph) and lazily evaluated. It limits the amount of data that is actually created because it allows the Spark framework to optimize the processing by analyzing many subsequent steps at once.

Each RDD is distributed across a Spark cluster; if your data source was HDFS, the initial partitioning will follow that of HDFS. In a sense, Spark's approach is similar to HBase. Regions in HBase are assigned to handle particular subsets of data based on their location in HDFS. The difference is that regions never actually store the data, while RDDs might, depending on the processing you apply to them. There are two types of processing that can be applied to RDDs, transformations, and actions. Most common examples are described in further sections of this chapter.

Not all data referenced through transformations and actions are persisted in memory. The user cannot distinguish between data cached in the memory and data read

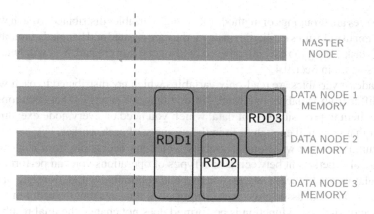

Fig. 9.1 Distribution of RDDs on a cluster

from the disk. It is made transparent by the framework. The only thing you might notice is a performance difference, but it would not be a repeatable behavior, because if you use a particular dataset more frequently, it will be kept cached.

In extreme cases, an RDD might be split between all nodes in the cluster like RDD A in Fig. 9.1. Most of the RDDs will occupy some parts of the cluster like RDD B and RDD C. You can choose to repartition RDD up or down to a chosen number of partitions;such operation might also reshuffle data across the cluster depending on the way it is performed.

In contrast to HDFS blocks, partitions of RDDs are not replicated. There is only one copy of each piece of data. Resilience, part of the name of RDDs, is not achieved through data redundancy. Instead, a minimal set of operations is determined for a lost partition based on the analysis of RDD's DAG. It means that in case a cluster node is lost, Spark will recalculate the missing data. Such an approach might be slower than simply recreating data from a copy. Nevertheless, it seems reasonable considering limited memory resources (comparing to disk space) and the fact that RDDs are ultimately not a persistent storage method.

DataFrames rely on RDDs and offer a higher level of abstraction. In many aspects, they are similar to data frames in Pandas and R. In Spark, they are created by enforcing a strict table-like data schema on an RDD. This enables higher level automatized processing routines, including the SQL-like interface. Not all data yields itself to be a DataFrame.

9.2.2 Other Data Structures

There are three more important data structures in Spark: closures, broadcast variables, and accumulators. Let us take a quick look at each of them.

Closures are groupings of methods (code) and variables distributed to each worker. You do not use closures explicitly; instead, they are created and handled automatically for each task during job execution based on the code you created. You will hear about closures again in Sect. 9.4.

Broadcast variables are read-only variables which are distributed to each worker. They differ from variables in closures in that they are persistent between jobs. You can use them to provision sets of data, which you need on every node executing the job, but that are small enough to fit in the memory of each worker.

Accumulators are write-only variables which are accessible from each worker. They are also persistent between tasks. Types of operations you can perform on the accumulators are limited. Only associative and commutative functions are allowed. If a function is both associative and commutative, it means that the order in which a sequence of such same function is performed does not change the final result.

9.3 Programming Model

In this section, I will discuss Spark's programming model. I will leave the description of the architecture for a further section. Basic programming operations are perfectly understandable and usable without fully understanding Spark's architecture, but the opposite is not necessarily true.

There are two types of operations in Spark's programming model—transformations and actions. Transformations are operations that create new RDDs. It might be natural to say that they alter the RDDs, but it would be incorrect. RDDs are immutable and each time you make any change a new RDD is created. Typical transformations include map, reduceByKey, but also more advanced operations like joins. Actions are operations that return data to the user; they include reduce, collect, and count.

Further, in this section, we will follow the basic example from the Hadoop 101 chapter. We will take all the basic operations from Hadoop and translate them to Spark. Do not assume that the operation with the same name has exactly the same results. Spark semantics are slightly different from that of Hadoop, despite similar terminology.

9.3.1 Data Ingestion

There are three general ways to create an RDD out of your data. It depends on where your data is currently. The fastest way is to covert an existing Python variable, using Spark's `parallelize` method. It is a convenient way during initial development but should be avoided for large datasets. You can see an example in Listing 9.1. First, a file is read into a variable, which is later converted into an RDD. Data does not have to necessarily come from a file. You can *parallelize* any variable, which is an iterable or a collection.

Listing 9.1 Loading data to RDD from a variable

```
1 hadoop_file = open('/dis_materials/hadoop_1m.txt')
2 hadoop_rdd = sc.parallelize(hadoop_file)
```

Another possibility is to load data from a file in a local file system, presented in Listing 9.2. This can be useful in many applications when data is not already present in a legacy file system. Loading large dataset this way might be slow because the load is not distributed. You will, however, not notice it right away due to lazy execution.

Listing 9.2 Loading data to RDD from local file system

```
1 hadoop_rdd = sc.textFile('file:///root/dis_materials/hadoop_1m.txt')
```

One of the most common ways to load data to an RDD is to read it from a file in HDFS, presented in Listing 9.3. In addition to HDFS, Spark supports also several other file systems, e.g., Amazon S3. It also supports databases, such as HBase or Cassandra. Such an approach to data loading is efficient because it can be done in a distributed fashion.

Listing 9.3 Loading data to RDD from HDFS

```
1 hadoop_rdd = sc.textFile('/dis_materials/hadoop_1m.txt')
```

9.3.2 Basic Actions—Count, Take, and Collect

The most important property of all actions is that they return value or values, which are presented to the user (that is you). It means that they finish a chain of lazily executed operations and cause actual data processing. Let us look at the three most common actions: count, take, and collect. They are all presented in Listing 9.4, and result of their execution on the example dataset is presented in Listing 9.5.

Listing 9.4 Basic actions in Spark—count, take, and collect

```
1 hadoop_rdd.count()
2 hadoop_rdd.take(5)
3 hadoop_rdd.collect()
```

The count does not take any arguments and returns cardinality of an RDD. In our case, the result should be 1000000. This action does not simply present the value, but it goes through the whole processing DAG to calculate the final amount of elements. It might take a while to get the result back, be patient.

Take returns first N elements, where N is the argument provides. In our case, we display five elements. These are always first elements, where order is determined by the arrangement of original data and subsequent processing. Unless you explicitly specify sort operation, data will not be sorted.

Listing 9.5 Sample result after applying actions on raw data

```
1 >>> hadoop_rdd.count()
2 INFO DAGScheduler: Job 7 finished: count at <stdin>:1, took 1.700608 s
3 1000000
4 >>>
5 >>> hadoop_rdd.take(5)
6 INFO DAGScheduler: Job 6 finished: runJob at PythonRDD.scala:393, took
      0.071443 s
7 [u'From general-return-2133-apmail-hadoop-general-archive=hadoop.apache.
      org@hadoop.apache.org Fri Oct 01 15:07:28 2010', u'Return-Path:
      <general-return-2133-apmail-hadoop-general-archive=hadoop.
      apache.org@hadoop.apache.org>', u'Delivered-To:
      apmail-hadoop-general-archive@minotaur.apache.org', u'Received: (qmail
      98655 invoked from network); 1 Oct 2010 15:07:28 -0000', u'Received: from
      unknown (HELO mail.apache.org) (140.211.11.3)']
```

In addition to their primary function, `Count` and `take` are very useful for testing your code. Because they are actions by calling them you force Spark to execute all the *lazy* code you specified earlier. I recommend to always call both of them. While `take` might be more informative, it, in fact, does not always result in processing the whole original dataset. It means that some issues might be left unnoticed. `Count`, on the other hand, by definition goes through all the data.

`Collect` returns all the results and, therefore, should be used with caution. It is not uncommon to operate on RDDs with millions or more elements. Trying to suddenly display all of them on the screen is not a good idea and is seldom necessary. Listing 9.5 omits results of `collect` and I do not recommend running it on this dataset. Feel free to test it on a smaller dataset later in this section.

9.3.3   Basic Transformations—Filter, Map, and reduceByKey

Now that we know how to extract key information and samples of data from RDDs, we can take a look into basic transformations. We can do it with a simple program we already know from earlier chapters. How different companies contribute to Hadoop development. We answer this question by counting the frequency of emails per domain. Implementation of this simple program uses all basic transformations: filter, map, and reduceByKey. It is presented in Listing 9.6.

`Filter` transformation applies a condition to each element (usually a line) of input. Only if the condition is met, then such line is output to a new RDD or next stage of processing. In this case, we are looking only for the line that starts with "From:", because such lines contain email of a sender. It is wise to use `filter` operation early in your program to reduce the amount of data considered as early as possible.

Listing 9.6 Basic transformations in Spark—filter, map, and reduceByKey

```
1 domain_counts = (hadoop_rdd
2                  .filter(lambda line: line.find("From:") == 0)
3                  .map(lambda line: (line[line.find("@")+1:line.find(">")],1))
4                  .reduceByKey(lambda a, b: a+b))
```

The next step is the map which extracts email domain from each line. It works exactly the same as in earlier Hadoop example. In fact, in Hadoop, it would be possible to connect filter and map operation because Hadoop's implementation of the map is not flexible. Spark's map always maps from 1 element to 1 element. Otherwise, it works the same as in Hadoop.

The last step is reduceByKey. The only thing it is doing is to perform a basic counting. But it might seem strange for two reasons. First, why is it not simply called *reduce*? Spark's regular reduce does not group data by key as it is standard in Hadoop. However, such grouping is a behavior we expect. Usually, reduceByKey is the best default replacement in Spark for reduce in Hadoop programs. Another operation groupByKey is sometimes considered to be of most direct correspondence to Hadoop's reduce, but it is slower than reduceByKey because it requires data shuffling and, therefore, it should be avoided.

The second reason, why is it not an action? In the end, we are used to *reduce* to finish data processing and output the results. The output of reduceByKey might be useful as input for further processing. The only way to enable such use is to have the output in a form of an RDD. Regular reduce is, in fact, an action as you can see in the next subsection.

Because all these operations are transformations, you have to use actions such as count or take to see the results. Look at Listing 9.7 to see an example.

Listing 9.7 Sample result after applying transformations

```
1 >>> domain_counts.count()
2 ...
3 INFO DAGScheduler: Job 9 finished: count at <stdin>:1, took 0.063379 s
4 269
5 >>>
6 >>> domain_counts.take(5)
7 ...
8 INFO DAGScheduler: Job 8 finished: runJob at PythonRDD.scala:393, took
    0.037368 s
9 [(u'nokia.com', 2), (u'sabalcore.com', 3), (u'cs.ucf.ed', 6),
    (u'cloudera.com', 442), (u'schiessle.org', 5)]
```

9.3.4 Other Operations—flatMap and Reduce

Regular map always returns one element of output for one element of input. Neither less nor more. We already learned about the filter to deal with cases when we might not want to return any output for some inputs. Another operation flatMap will deal with the opposite case when we want to return more than one output for some inputs.

Such a case is quite common. In fact, flatMap is used as frequently as a map. Their use simply depends on the expected structure of the output. To show the difference between them, let us count domain endings—such as .com or .org—in our dataset. At this stage, we can focus on data preparation. We want to extract all the domain endings from hadoop_rdd. In the first three lines of Listing 9.8, you can see an example code using map. It in principle works correctly, but can the output it produces might be inconvenient to work with? Because each domain might contain more than one ending, e.g., com and au, or co and uk, the output is a list of lists. You can see it in Listing 9.9, line 13. It would be easier to work with one simple list.

Listing 9.8 Difference between map and flatMap in Spark

```
1  domain_endings_map = (hadoop_rdd
2                        .filter(lambda line: line.find("From:") == 0)
3                        .map(lambda line:
                             line[line.find("@")+1:line.find(">")].split('.')[1:]))
4
5  domain_endings_flat = (hadoop_rdd
6                         .filter(lambda line: line.find("From:") == 0)
7                         .flatMap(lambda line:
                              line[line.find("@")+1:line.find(">")].split('.')[1:]))
```

In such cases, flatMap can produce exactly the output we expect. FlatMap simply flattens the list of lists into one list, as you can see in Listing 9.9, line 18. Such an approach works not only on lists but also other collections, e.g., tuples. Try to change the map to flatMap in the original example from Listing 9.6 and observe the difference in results.

Listing 9.9 Difference between map and flatMap in Spark—Results

```
1  >>> domain_endings_map.count()
2  ...
3  INFO DAGScheduler: Job 2 finished: count at <stdin>:1, took 2.001068 s
4  11424
5  >>> domain_enginds_flat.count()
6  ...
7  INFO DAGScheduler: Job 3 finished: count at <stdin>:1, took 2.063895 s
8  11583
9  >>>
10 >>> domain_endings_map.take(5)
11 ...
12 INFO DAGScheduler: Job 4 finished: runJob at PythonRDD.scala:393, took
       0.034376 s
13 [[u'com'], [u'com'], [u'com', u'au'], [u'com'], [u'com']]
14 >>>
15 >>> domain_endings_flat.take(5)
16 ...
17 INFO DAGScheduler: Job 5 finished: runJob at PythonRDD.scala:393, took
       0.029242 s
18 [u'com', u'com', u'com', u'au', u'com']
```

To actually calculate amount of different domain endings, we still have to extract only the unique endings from our current dataset and then count them. However, with data properly prepared with flatMap, it is an easy task. I encourage you to do it.

In contrast to flatMap, reduce is an action. Moreover, in further contrast, to reduce in Hadoop is does not group elements by key. That is what reduceByKey does, which is a transformation. To see how the reduce works, we can implement a simple counting algorithm; it is presented in Listing 9.10. In map, we simply output 1 for each element and in reduce we add these ones. In Listing 9.11, you can notice that we do not have to use `take` to display the result. Reduce returns data directly because it is an action.

Listing 9.10 Reduce in Spark

```
1 domain_counts.map(lambda a: 1).reduce(lambda a, b: a+b)
```

Listing 9.11 Reduce in Spark—Results

```
1 ...
2 INFO DAGScheduler: Job 10 finished: reduce at <stdin>:1, took 0.040092 s
3 269
```

Reduce is less frequently used than reduceByKey. Nevertheless, it is a useful operation whenever you need to perform an aggregation that ignores keys and you want simply to get the result instead of creating another RDD.

9.4 Architecture

Spark uses two types of nodes to organize the work on the cluster, driver node, and worker nodes. Such architecture is similar to what you have seen in Hadoop and what is common in many distributed processing systems.

There are two types of nodes in Spark architecture, driver node, and worker nodes. You can notice a similarity to Hadoop and, in fact, many other distributed processing systems. The purpose of the drive is to control work on a cluster, while works perform the processing tasks assigned to them by the driver. The driver holds several modules and services, SparkContext, schedulers (DAGScheduler and Task Scheduler), several management services, and other elements like WebUI. SparkContext is a process that is a basis for your communication with the cluster. Each user accessing Spark cluster gets their own context.

In Fig. 9.2, you can see a top-level view over main Spark elements. Each worker consists of two main parts, executor and cache. The executor can run multiple tasks at a time. Cache is an in-memory structure that maintains parts of RDDs for immediate access. Data comes either directly from persistent storage or is a result of processing. Not all RDDs are materialized in cache all the time, for example, due to insufficient memory. If such RDD is requested, it will be recreated; this process is transparent to the programmer.

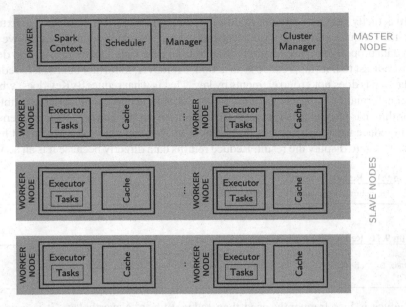

Fig. 9.2 Spark architecture

SparkContext communicates with worker nodes using a cluster manager. There are several different managers that can be used. Spark's standalone manager provides a simple way to deploy spark installation without advanced management software. Mesos is a cluster manager based on a concept of kernels and it originates from the same source as the Spark itself. YARN is a cluster manager originating from the Apache Hadoop project.

A job starts when an action is executed on an RDD. The execution plan is based on RDDs DAG. This means that the combination of transformations you applied earlier to the RDD will naturally influence the executions plan. The order of transformations is not without importance. Spark will try to optimize some typical scenarios, but ultimately it does not understand your code so it cannot optimize everything.

Each job consists of stages. The stage is characterized by execution of the same code but on different subsets of data localized on different workers. Each such execution is called a task. SparkContext sends tasks to executors to run. The code and variables passed to each executor are called a closure.

Each stage can contain many transformations, as long as data does not have to be shuffled. Transformations like map and filter do not cause shuffling and they are called *narrow transformations*. On the other hand, transformations like groupByKey or reduceByKey do cause shuffling and they are called *wide transformations*.

Shuffling in Spark is similar to a shuffle in Hadoop or MapReduce in general, but in contrast, it is not limited to the phase between map and reduce. In fact, Spark organizes shuffle using tasks called map and reduce, but they are not the same as map transformation and reduce action you can execute as a user. Data for shuffle is written to the local disk and then fetched over the network by the next stage.

Various combinations of transformations and actions can lead to the same result, but the execution plan might differ. It means that performance might differ too. Spark Catalyst will attempt to optimize the execution plan, but it cannot cover all the possible scenarios. Therefore, the way you combine transformations and actions is not without importance for the final performance.

Spark uses JVMs to execute the code, but in contrast to Hadoop, it does not restart JVMs for each task. For iterative algorithms, it can result in more than an order of magnitude improvements. Spark uses, what is called, a lazy execution. It can be tricky, especially for debugging, in the beginning. Lazy execution waits with the actual execution of any code until a result of this execution is actually requested. In Spark context it requesting, the result is synonymous with executing an action.

As a result, any bugs in the code will not be immediately visible. It can be confusing at first and might remain a challenge later on too. At the same time, lazy execution gives better-optimized code, because Spark Catalyst gets on a better overview of the whole data pipeline to optimize.

During code development, it might be advisable to temporarily add test actions. These will be removed later but will help you to ensure correctness of each stage of the code.

In addition, there are multiple libraries built on top of basic Spark Architecture. Four most important ones are SQL—providing higher level access method to Spark data, MLlib—implementing scalable machine learning algorithms, GraphX—implementing graph algorithms, and Streaming—extending Spark from batch to stream processing. I introduce you to SparkSQL and I hope you will explore the other libraries yourself.

9.5 SparkSQL

In this section, I will show you how to recreate the example you did in Hive at the beginning of this book, but instead, we will use SparkSQL.

We first load the necessary libraries, start Spark's SQLContext, and read the input file. You can see it in Listing 9.12, in lines 1–3. We use a CSV file, which in principle could be imported directly into SparkSQL. However, it would require additional libraries at the moment. The file we use here is very simple so it is easier to process it as a regular text file, lines 5–7. Finally, in lines 9–10, we create a new table in sqlContext based on the RDD emails we created and we register table name.

Listing 9.12 Register table for in SparkSQL

```
1 from pyspark.sql import SQLContext, Row
2 sqlContext = SQLContext(sc)
3 hadoop_csv_rdd = sc.textFile('/dis_materials/hadoop_1m.csv')
4
```

```
5 emails = (hadoop_csv_rdd
6          .map(lambda l: l.split(","))
7          .map(lambda p: Row(id=p[0], list=p[1], date1=p[2], date2=p[3],
                  email=p[4], subject=p[5])))
8
9 schemaEmails = sqlContext.createDataFrame(emails)
10 schemaEmails.registerTempTable("emails")
```

Now we can specify our query in Listing 9.13. This is in principle the same query you have used for Hive example. Usually, only minor adjustments are necessary when you migrate typically queries from Hive or SQL databases.

Five first example results are presented in Listing 9.14.

Listing 9.13 Execute query in SparkSQL

```
1 lists = sqlContext.sql("SELECT substr(email, locate('@', email)+1) AS domain,
     count(substr(email, locate('@', email)+1)) AS count FROM emails GROUP BY
     substr(email,locate('@', email)+1)")
```

Listing 9.14 Execute query in SparkSQL—Results

```
1 ...
2 INFO DAGScheduler: Job 3 finished: take at <stdin>:1, took 0.050643 s
3 [Row(domain=u'alibaba-inc.com', count=4), Row(domain=u'physics.ucsd.edu',
     count=2), Row(domain=u'Boudnik.org', count=1),
     Row(domain=u'teamrubber.com', count=1), Row(domain=u'omniti.com', count=7)]
```

It is important to notice that the result of executing a SparkSQL query is, in fact, an RDD. It means you can use such result for further processing with the regular map and reduceByKey functions. In Listing 9.15, you invoke the *RDD* underneath the *DataFrame* that holds results of the SparkSQL query and remove the domain ending, such as *.com* from each result. A few first lines from example output are seen in Listing 9.16.

Listing 9.15 Execute map operation on output from SparkSQL

```
1 lists.rdd.map(lambda line: line.domain[:-4])
```

Listing 9.16 Execute map operation on output from SparkSQL—Results

```
1 [u'cn.ibm', u'odman.int.dev', u'deri', u'yieldex', u'oskarsso', u'rosenlaw',
     u'adobe'...
```

9.6 Exercises

Exercise 9.1 Extend the example from Listing 9.6 to perform all operations from Sect. 7.1 in Chap. 7.

Exercise 9.2 Finish the example from Listing 9.8 to actually count the unique domain endings.

Exercise 9.3   Recreate all remaining patterns from Chap. 7 in Spark.

Exercise 9.4   Consider use cases for Hadoop and Spark. Is Spark a replacement for Hadoop? Should they coexist?

Printed in the United States
By Bookmasters